Lighting Controls Handbook

Lighting Controls Handbook

Craig DiLouie

River Publishers

Routledge
Taylor & Francis Group
LONDON AND NEW YORK

Published 2020 by River Publishers

River Publishers

Alsbjergvej 10, 9260 Gistrup, Denmark

www.riverpublishers.com

Distributed exclusively by Routledge

4 Park Square, Milton Park, Abingdon, Oxon OX14 4RN

605 Third Avenue, New York, NY 10017, USA

Library of Congress Cataloging-in-Publication Data

DiLouie, Craig, 1967-
    Lighting controls handbook / Craig DiLouie.
      p. cm.
    Includes index.
    ISBN 0-88173-573-6 (alk. paper) -- ISBN 978-8-7702-2269-3 (electronic) -- ISBN
978-1-4200-6921-1 (Taylor & Francis : alk. paper)
      1. Electric lighting--Automatic control--Handbooks, manuals, etc. 2.
Electric power--Conservation--Handbooks, manuals, etc. I. Title.

    TK4169.D57 2006
    621.32--dc22

                                                                        2007030706

Lighting controls handbook / Craig DiLouie.
First published by Fairmont Press in 2008.

Routledge is an imprint of the Taylor & Francis Group, an informa business

0-88173-573-6 (The Fairmont Press, Inc.)
978-1-4200-6921-1 (print)
978-8-7702-2269-3 (online)
978-1-0031-5140-1 (ebook master)

Table of Contents

Foreword

Lighting controls play a critical role in fluorescent lighting systems, providing the function of 1) turning the lights on and off using a switch; and/or 2) adjusting light output up and down using a dimmer.

In recent decades, technological development has increasingly automated these functions and allowed integration of devices into larger, more sophisticated systems. The result is significantly expanding energy-saving opportunities, flexibility, reliability and interoperability between devices from different manufacturers.

One thing remains the same: A good lighting design includes a good controls design. The goal of an effective control system is to supporting the lighting application goals, which often translates to eliminating energy waste while providing a productive visual environment.

The goal of adjusting the state of a fluorescent lighting system may be to achieve energy savings, support occupant visual needs and preferences, or a combination of the two. These actions may be taken by the system designer for the purposes of saving energy, adding value to a project, satisfying code requirements, earning LEED points and/or achieving a range of other goals.

The Lighting Controls Handbook provides detailed information about fluorescent switching and dimming systems in commercial buildings, covering technology, design, application, strategies, typical savings, commercial energy code compliance, research, commissioning and troubleshooting.

This book is intended for electrical engineers, energy managers, building managers, lighting designers, consultants and other electrical industry professionals interested in lighting controls as a means of energy savings and supporting visual needs.

Automatic lighting controls are required by prevailing energy codes and can be used to save energy, support visual needs and other purposes in a broad range of applications. Photos courtesy of Leviton (top) and Advance (bottom).

Introduction to Lighting Controls

THE LIGHTING CONTROL SYSTEM

Lighting control systems contain three components linked by communication wiring, which is used to transmit control signals, and power wiring, which supplies power, as shown in Table 1.

Table 1. Basic function of a lighting control.

Component	Sensing Device →	Logic Circuit →	Power Controller
Function	Provides information to logic circuit	Decides whether to supply lighting, and how much	Changes the output of the lighting system

We can therefore view a lighting control system or device as an apparatus that 1) receives information, 2) decides what to do with that information, and 3) changes the operation of the lighting system. In other words, we can look at lighting control devices based on inputs and outputs. Three examples are shown in Table 2.

Figure 1 shows an example of a robust lighting control system with a control station, occupancy sensor, photosensor and time switch or centralized switching system providing a variety of inputs to the master lighting controller. The lighting controller can be a switching panel, dimming panel or both linked together. The controller in turn controls the lighting load with a variety of outputs based on decisions made by the logic circuits. Since different control strategies may have overlapping device requirements, control synergies can be gained by building a system of simple components.

PURPOSE OF LIGHTING CONTROLS

In many applications, the overall purpose of the lighting control system is to eliminate waste while supporting a productive

Table 2. Examples of functionality of various control devices expressed as inputs and outputs.

Control	Input	Decision-making	Output
Occupancy sensor	Sensor detects presence or absence of people	Decide whether to turn on or shut off lights	Sends signal to relay which closes or opens circuit
Control station and dimming panel	User presses button to recall preset scene	Control station recalls scene from memory and sends signal to dimmer at dimming panel	Dimmer adjusts light output to desired level
Dimmable ballast	Controller provides signal to dim	Ballast is instructed to dim, and by how much	Ballast alters the current to the lamps, dimming them

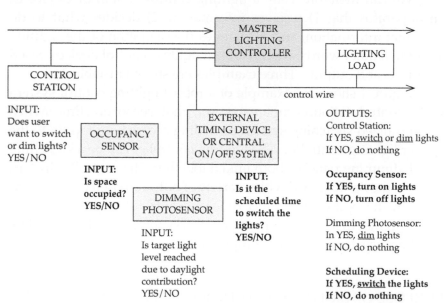

Figure 1. Centralized programmable control system with inputs from control accessories.

visual environment. This entails:

- Providing the right amount of light
- Providing this light where it's needed
- Providing this light when it's needed

The Right Amount of Light...

Control systems provide the right amount of light. This lighting decision is based on the type of tasks being performed in the space. Lighting controls support this goal in two ways.

Lighting controls provide flexibility in adapting the lighting system to different uses. For example, a school auditorium, which is home to a diverse range of activities, would need different light levels for these activities.

Lighting controls provide the ability for users to adjust light levels based on changing needs or individual preference, either through dimming or through bi- or multi-level switching. Dimming provides the greatest amount of flexibility in light level adjustment.

By enabling the lighting system to deliver the right amount of light to the task, the control system can eliminate energy waste while supporting a productive visual environment.

...Where It's Needed...

Lighting controls support the lighting system putting light where it's needed. This entails establishing control zones, which is a light fixture or group of fixtures controlled simultaneously as a single entity by a single controller. (Control zones may be called "channels" by manufacturers in their guides and equipment specifications.)

Zones are typically established based on types of tasks to be lighted, lighting schedules, types of lighting systems, architectural finishes/furnishings, and daylight availability. Zones should be established, if possible, not only on the immediate use of the space but also anticipated future uses.

Zones are often limited to the fixtures on one circuit or sub-

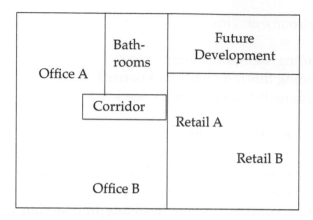

Figure 2. Lighting control zones should be established, if possible, not only on the immediate use of the space but also anticipated future uses. Source: U.S. Department of Energy (DoE).

circuit, or switch-leg, but similar fixtures on multiple circuits up to large loads can be zoned to be controlled by a single controller as well. Establishing smaller zones typically increases control accuracy, flexibility and energy savings, but also increases initial cost. For example, a control system can turn the lights on automatically when a person enters a building during non-operating hours. Only the areas to be used should be lighted, however, and not the entire floor. A zone can also be as small as single ballast or light fixture, which enables the greatest amount of control resolution. For example, each user in an open office can be given capability via PC or handheld remote to dim his or her own lighting to personal preference.

The right approach to control zones will depend on the application and use of controls. Generally, smaller zones are used for bi-level switching based on manual control, multi-level switching based on daylight availability, and occupancy-based controls to prevent a person entering a space after hours from causing the lighting in a larger area to activate than necessary. Larger zones are generally used for scheduling.

... And When It's Needed

An effective control system ensures that the lighting system operates—and consumes energy, which costs the owner money—only when it's needed. Determining when the lighting system should be operating depends on how the space is occupied. This will entail whether a time-based or a threshold event should be the deciding factor in whether the lights should be turned on or shut off.

If occupancy is predictable, a time-based strategy can be considered. For example, a switching system can be scheduled to automatically shut off the lights by area, by floor or in an entire building if a building's occupancy is predictable.

If occupancy is not predictable, a threshold-event-based strategy can be considered. For example, occupancy sensors can be used to automatically turn on and shut off lights in areas depending on whether the sensor detects the presence or absence of people in the monitored area.

By ensuring the lighting system provides light only when it's needed, the control system can significantly reduce wasted energy and generate utility cost savings for the owner.

Energy Management

Advanced lighting control devices and systems can be used to reduce ongoing costs for the owner and thereby increase profitability and competitiveness. According to the New Buildings Institute, lighting controls can reduce lighting energy consumption by 50 percent in existing buildings and by at least 35 percent in new construction.

- *Lighting energy:* Controls can reduce the amount of power drawn by the lighting system during operation and also the number of operating hours, thereby reducing utility energy charges.

- *Lighting demand:* Controls can reduce the amount of power drawn by the lighting system, reducing utility demand charges—particularly during peak demand periods, when demand charges are highest.

Depending on the application, these cost savings can produce a short payback and a high rate of return for the investment in the new controls. In new construction, the rate of return is often higher because only the premium, not the total installed cost, will be recouped before positive cash flow is realized.

Performing an Economic Analysis

You'll need to present the conceptual control design (i.e., color-coded diagram or list of dimming and other control ideas) to the client and/or the owners, and they are bound to wonder about the bottom line. Perform a simple, rough economic analysis that involves a walk-through to determine baseline equipment and energy usage. Note that 30-45 percent of a building's electricity bill is typically lighting, and 30-35 percent of the cost of a building is for mechanical systems and envelope architecture. Lighting controls can contribute significantly to cost savings in these areas. For example, controls can reduce cooling equipment costs by \$0.10 to \$0.15 per square foot, by some estimates.

One way to perform an economic analysis for lighting controls is using an "effective energy charge" (EEC), which is a \$/kWh figure. To determine this value, use the below formula:

Value of Entire Electric Bill (\$) ÷ Total Amount of Energy Used During Bill Period (kWh)

You can then determine an approximate estimate of potential annual dollar savings/controlled light fixture using this formula:

Savings/Year (\$) = ([IW x IOH x IPF] – [FW x FOH x FPF]) * (EEC) * (1kW/1000W)

Where ...

IW	=	Initial watts (W), baseline input watts for controlled fixture(s)
IOH	=	Initial operating hours/year for controlled fixture(s)
IPF	=	Initial power fraction, baseline power fraction for the fixture(s) (use 1 for full power)
FW	=	Final watts (W), input watts for controlled fixture(s) after controls upgrade
FOH	=	Final operating hours/year for controlled fixture(s) after controls upgrade
FPF	=	Final power fraction, power fraction for the controlled fixture(s) after controls upgrade (e.g., if fixture will be switched or dimmed to annual average of 75 percent, use 0.75)
EEC	=	Effective energy charge (\$/kWh)

Visual Needs

The project may be driven by business benefits other than energy savings by providing increased performance and flexibility:

- Adapt the lighting for multiple uses of a space, such as a conference room or gymnasium

- Adapt the lighting to evolving space needs resulting from employee churn and office strategies such as hoteling and hot-desking

- Mood-setting for restaurants and similar applications

- Increasing worker satisfaction by providing personal control of their lighting systems in office and other environments

- Enhanced aesthetics and image, greater space marketability, and pollution prevention

These business benefits are often more difficult to calculate than energy savings, but can tangibly contribute to the bottom line.

Studies, for example, have shown that personal lighting control can increase worker satisfaction, a major contributor to productivity, while providing energy savings. According to the Building Owners and Managers Association (BOMA), energy costs run about $2/sq.ft. in a typical commercial building while worker salaries and benefits can run to $130/sq.ft. or more. While reducing energy costs by a large percentage can be profitable, increasing productivity by even a very small percentage can be much more profitable.

Energy Codes

Lighting automation is now mandated in most of the United States. With the adoption of the ASHRAE/IES 90.1-1999 model energy code by DoE as the minimum national standard, a majority of state energy codes require automatic shut-off of all lighting in commercial buildings greater than 5,000 sq.ft. in size, with few exceptions. Automatic shut-off can be provided by occupancy sensors and programmable time scheduling devices. Some state-specific codes, such as California's Title 24, are even stricter. The 2005 Title 24 enacted in October 2005, for example, requires daylighting controls in some spaces as well as commissioning for control systems. For more information about U.S. energy codes, visit www. energycodes.gov. For more information specifically about lighting control code requirements, see Chapter 9.

Determining a Lighting Control Strategy
The below questions can help narrow down selection of an appropriate control strategy:

• What are the load and space characteristics?

• What are the project goals for the lighting system?

• What is required by energy, building and electrical codes?

• Should the load be switched or dimmed?

• What degree of automation is required?

• Should the lighting system be controlled locally, from a central location, or both?

• What degree of control accuracy is required?

• What is the target value level—the balance between performance/capabilities and cost?

What are the application goals?
Table 3 shows three studies that suggest the most important reasons why advanced lighting controls are specified.

Table 3. Why are advanced lighting controls specified?

	2003 Ducker Research/Watt Stopper Lighting Automation Study	2004-2005 ZING Communications/ LCA Dimming Study	2005 Square D Bulls Eye Study
Methodology	Telephone interviews of 158 facility managers, electrical engineers and architects	Email survey to 4,317 lighting designers, architects, engineers, distributors and contractors with 6.7 percent response	Direct mail survey
Study Focus	Automatic Switching	Dimming	Lighting Controls
Research Question	What are the top five factors driving the use of automatic lighting controls?	What are the top five reasons for specifying dimming systems?	What are the most important benefits of lighting control?
Average Respondent Answer	1. Increasing energy savings 2. Complying with owner requests 3. Compliance with state and national energy codes 4. Providing occupant control capability 5. Obtaining utility rebates and incentives	1. Ability to light space for different uses (flexibility) 2. Client request 3. Energy savings 4. Add value to the design 5. Mood setting	1. Reduce energy costs 2. Worker safety 3. Occupant convenience 4. Prolong equipment life 5. Meet state energy codes

Switching or Dimming?

The first primary decision after defining the load and the application goals is whether to switch or dim the load. Switching and dimming are stand-alone strategies but are often used in the same facility, and may be integrated in the same control system.

Table 4 provides a general comparison between switching and dimming.

Local or Central Control?

Table 5 provides a general comparison of localized and centralized approach to lighting control. Often, a minimum starting point will be established by the applicable energy code.

What Degree of Automation is Required?

Manual lighting controls range from a single switch to a bank of switches and dimmers that are actuated by toggles, rotary knobs, push

Table 4. General comparison of switching and dimming.

Method	Switching	Dimming
Primary Use	Energy management	Visual needs
Basic Function	Turn lights on or off	Change light output with smooth transitions between light levels
Benefits	Utility cost savings	Occupant satisfaction, flexibility, utility cost savings
Advantages	Relatively inexpensive and simple to commission; limited selection of light levels (bi-level switching)	Can set light output at any level within available range, greater user acceptance due to smooth transitions between light levels
Disadvantages	Lower user acceptance in occupied spaces with stationary tasks due to abrupt, noticeable changes in light level (automatic multi-level switching); limited energy savings (manual switching)	Higher installed cost, and can require more sophisticated commissioning depending on the size and complexity of the system

Table 5. General comparison of localized and centralized control.

Control Method	
Divide the building into series of control zones, each zone constituting a lighting load controlled by a single controller	
Localized	Centralized
Each zone operated by its own point of control independently of other zones	All zones operated by single point of control
Lower cost, less sophisticated commissioning	Greater capabilities, flexibility, potential cost savings

The control scheme can be designed with local systems and a centralized system working together as layers. Both local and centralized systems can be integrated into building automation systems for control of lighting and HVAC.

buttons, remote control, and other means. Manual controls can be cost-effective options for small-scale situations. However, as the size of the lighting system grows, manual controls lose their cost-effectiveness. In addition, manual controls often waste energy because the decision to shut off the lights when they are not needed is based entirely on human initiative.

Table 6 provides a list of common automatic and manual control strategies.

Table 6. Common automatic and manual control strategies.

Method	Strategy	Manual vs. Automatic
Switching	Occupancy sensors	Turn the lights on or off **automatically** based on whether space is occupied
	Scheduled automatic shut-off at end of workday switching panels, time-clocks or building automation system	Turn selected lights on or off **automatically** based on schedule when space is predictably unoccupied
	Scheduled automatic shut-off of select loads (bi-level switching) during peak demand periods, time-clocks or building automation system	Turn off one or two lamps in each fixture or checkerboard fixtures **automatically** for load shedding during peak demand periods
	Bi-level switching using wall switches controlling lighting system layered as two separate circuits	Turn selected circuit on and off **manually** to achieve on status, 50 percent light level, and off
	Multilevel switching using photosensor and low-voltage relay	Turn the lights off **automatically** based on available ambient daylight
Dimming	Dimming control of smaller loads using wall-box and remote dimmers	Adjust light output **manually** based on space need or personal preference
	Dimming control of larger loads using control stations and dimming panels	Adjust light output **manually** based on space need or personal preference
	Daylight harvesting using photosensor, controller and dimmable ballast	Adjust light output **automatically** to maintain target light level as daylight enters space
	Adaptive compensation using dimming panels and scheduling device such as time-clock	Adjust light output **automatically** to provide lower light levels at night based on studies about human lighting preferences
	Peak shaving and load shedding using dimming panels and scheduling or control device	Adjust light output **automatically** during peak demand periods and / or **manually** based on utility request to curtail load

What Degree of Control Accuracy is Required?

A key step in designing a lighting control system is to determine the degree of control over the lighting system, which means breaking the load up into zones. Establishing smaller zones increases control accuracy and flexibility but also increases cost (see Figure 3). In some cases, energy codes will establish minimum control zones.

Get Started

By determining the most appropriate control strategy for the application using these simple guidelines, one can specify control systems that provide the right amount of light where it's needed, and when it's needed—thereby optimizing operating costs and user satisfaction.

Table 7. Common lighting control strategies, indicating whether, or to what extent, each is a dimming vs. switching strategy, local vs. centralized strategy, and manual vs. automatic strategy.

Occupancy sensors

Dimming		▓	Switching
Local	▓		Centralized
Manual		▓	Automatic

Scheduling (automatic shut-off)

Dimming		▓	Switching
Local		▓	Centralized
Manual		▓	Automatic

Bi-Level/Multilevel switching

Dimming		▓	Switching
Local	▓		Centralized
Manual	▓		Automatic

Task Tuning (Personal Dimming)

Dimming	▓		Switching
Local	▓		Centralized
Manual	▓		Automatic

Daylight Dimming

Dimming	▓		Switching
Local	▓		Centralized
Manual		▓	Automatic

Scheduled Dimming

Dimming	▓		Switching
Local		▓	Centralized
Manual		▓	Automatic

↑ Smaller zones

Control accuracy

Flexibility

Complexity

Cost

Larger zones ↓

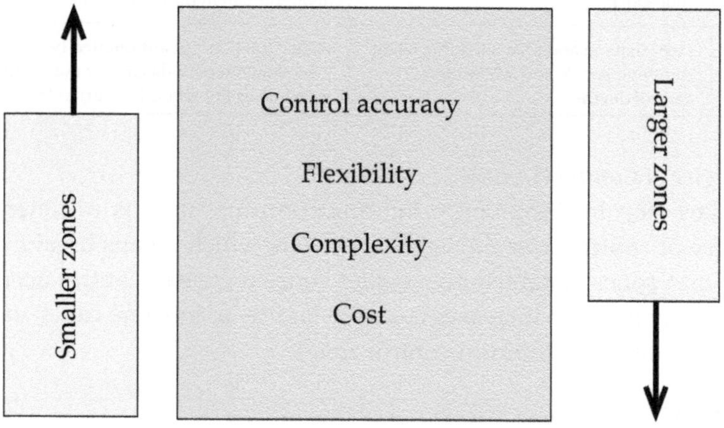

Figure 3. Possible tradeoffs to consider when sizing a control zone.

Part I

Switching Fluorescent Systems

Chapter 1

Occupancy Sensors

While a building's lighting systems are typically activated in the morning and stay on all day, buildings typically experience a daily lighting usage profile that peaks in the afternoon and declines until the end of the day. In the morning and after normal hours, many lights are operating but may not be needed. Studies suggest that workers are not in their offices 30-70 percent of the time during work hours.

Occupancy sensors (see Figure 1-1) are lighting control devices that detect when a space is occupied or unoccupied and turn the lights on or off automatically after a short period of time to save energy.

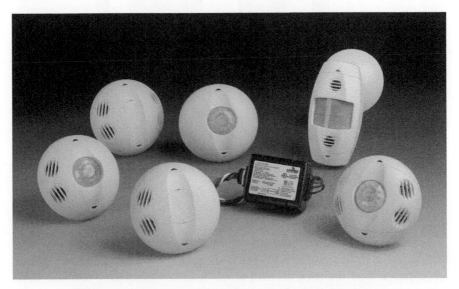

Figure 1-1. Occupancy sensors. Courtesy of Leviton.

By turning off the lights when people are not using the lighting in a given space, the owner realizes energy cost savings, particularly in spaces that are occupied intermittently throughout the day (see Figure 1-2). Energy savings are variable based on application characteristics, but estimate range from a conservative 30 percent to 90 percent at the high end.

In addition, occupancy sensors have the potential to provide low-cost security for indoor and outdoor spaces by indicating that a space is occupied, and can minimize light pollution by reducing the use of both indoor and outdoor lighting.

Typical applications include private offices, restrooms, classrooms, conference rooms, break rooms, etc. Occupancy sensors are ideally suited for spaces:

- In which the lighting is not required to be operating all day for safety or security reasons. For example, occupancy sensors are not recommended for public spaces such as hallways and

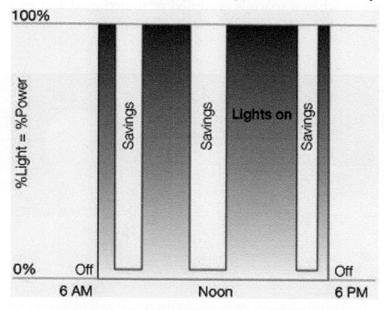

Figure 1-2. Studies suggest that office workers are not in their offices 30-70 percent of the time during work hours. Occupancy sensors save energy by turning off the lights when a space is unoccupied. Graphic courtesy of Lutron Electronics.

lobbies, where the lights must remain on even when the space is unoccupied.

• That are intermittently occupied throughout the day or are otherwise left vacant for significant portions of the day, and where occupancy is less predictable.

• Smaller projects.

• Projects requiring more granular control.

Due to their relative simplicity and high energy savings, occupancy sensors are rapidly becoming a standard feature in new buildings. But perhaps the most significant driver is energy codes. With varying exceptions and requirements, the 2003 and 2006 International Energy Conservation Code (IECC) and ASHRAE/IESNA Standard 90.1-1999, -2001 and -2004 all require interior and exterior automatic lighting shutoff controls as well as manual or automatic controls in interior enclosed spaces. By extension, it is also a prerequisite for earning Leadership in Energy & Environmental Design (LEED) green rating system points or the Energy Policy Act of 2005's accelerated tax deduction of up to $0.60/sq.ft. encouraging investment in efficient lighting, as these programs are based on meeting/exceeding the minimum requirements of Standard 90.1-2004 and -2001 respectively. Occupancy sensors are recognized as an acceptable automatic shutoff strategy in energy codes.

This chapter describes occupancy sensor technology, application and commissioning.

TECHNOLOGY

This section describes how occupancy sensors operate, major specifiable performance parameters, technologies, mounting configurations, special features, trial installations, and using light loggers to project energy savings.

Theory of Operation

Occupancy sensors are switching devices that are typically comprised of three components:

- Motion detector
- Power supply
- Relay switch

The relay switch and power supply are often housed in the same unit, which may be called a power-pack or switch-pack. The relay switch closes or opens the circuit to the connected lighting load, turning it on or shutting it off. The power supply's transformer converts 120V or 277V AC line voltage to AC low voltage required to power the sensor. See Figure 1-3.

These two units typically communicate via Class II low-voltage wiring. While most occupancy sensors are low-voltage, some are line-voltage. These sensors do not use a power-pack. They are suitable for applications where there is no plenum or junction boxes are hard to access. Be aware that line-voltage ceiling sensors may only have the ability to switch about one-third the load of a ceiling sensor that uses a power pack.

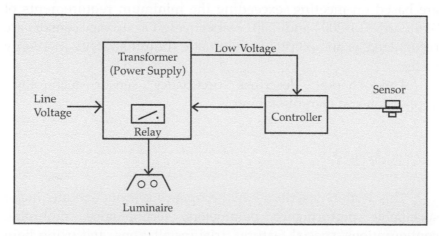

Figure 1-3. Components of typical occupancy sensor-based switching. Source: New Buildings Institute.

Energy Savings

Depending on the characteristics of the space to be controlled, energy savings as high as 90 percent can be realized through use of occupancy sensors, according to the U.S. Environmental Protection Agency (see Table 1-1).

Table 1-1. Typical energy savings, by application, for occupancy sensors. Source: U.S. Environmental Protection Agency.

Occupancy area	Energy Savings
Private office	13-50%
Classroom	40-46%
Conference room	22-65%
Restrooms	30-90%
Corridors	30-80%
Storage areas	45-80%

In 1997, researchers studied energy savings potential for occupancy sensors in buildings in 24 states representing a cross-section of commercial building types. The study monitored occupancy and the number of hours the lights were on in 158 rooms, including 37 private offices, 42 restrooms, 35 classrooms, 33 conference rooms and 11 break rooms. Potential energy savings for these spaces types are shown in Table 1-2.

Sensor Variables

Technology: Choose method of motion detection that will best meet the application need.

Sensitivity: Determine how sensitive the sensor should be to movement so that it effectively detects minor and major motion without nuisance switching (false-on/off).

Coverage area: Specify range (ft.) and coverage area (sq.ft.) for the motion detector based on the desired level of sensitivity.

Mounting: Locate the sensor for maximum effect.

Time delay: Determine how long the sensor should wait before turning out the lights when the space is unoccupied to be convenient

Table 1-2. Potential energy savings as revealed in sensor research study. Source: Maniccia, D. et al., "An Analysis of the Energy and Cost Savings Potential of Occupancy Sensors for Commercial Lighting Systems," Proceedings of the 2000 Annual Conference of the Illuminating Engineering Society of North America.

Space Type	Savings Potential All Hours	Savings Potential Normal Hours	Savings Potential After Hours
Restroom	60%	18%	42%
Conference room	50%	27%	23%
Private office	38%	25%	13%
Break room	29%	14%	15%
Classroom	55%	23%	35%

for users but also maximize energy savings.

Cut off: Determine whether the occupancy sensor's coverage area must be restricted so that it will not monitor adjacent areas that should not be monitored (such as a hallway outside a controlled private office).

Special features: Specify special features for the sensor based on available offerings from manufacturer and application need.

Sensor Technology

Occupancy sensors detect the presence or absence of people using one or a combination of several methods. The most popular methods are passive infrared (PIR) and ultrasonic. Dual-technology sensors use both methods.

Each method has advantages and disadvantages that make it more suitable for some applications than others.

PIR sensors: PIR (passive infrared) occupancy sensors (see Figure 1-4) sense the difference in heat emitted by humans in motion from that of the background space. These sensors detect motion within a field of view that requires a line of sight; they cannot "see" through obstacles.

The sensor's lens defines its coverage area as a series of fan-shaped coverage zones, with small gaps in between, and is most sensitive to motion that occurs between each zone (lateral to the sensor) (this can be seen later in this chapter, in Figure 1-7). The farther a person is from the sensor, the wider the gaps between these zones become, which decreases sensitivity proportional to distance and can result in nuisance switching (false-off).

Most PIR sensors are sensitive to full body movement up to about 40 ft. but are sensitive to hand movement, which is more discrete, up to about 15 ft.

Ultrasonic sensors: Ultrasonic occupancy sensors (see Figure 1-5) utilize the Doppler principle to detect occupancy through emitting an ultrasonic high-frequency signal throughout a space, sense the frequency of the reflected signal, and interpret change in frequency as motion in the space.

These sensors do not require a direct line of sight and instead can "see" around objects such as partitions. However, in open office areas, particularly spaces with fabric-covered partitions, direct line of sight between the sensor and the occupant may be required for reliable detection.

Unlike the lens of a PIR sensor, the ultrasonic sensor's transducers do not include gaps between discrete coverage zones making up the field of view and so can be sensitive to hand motion at distance up to 25 ft. However, the sensitivity of ultrasonic sensors makes them vulnerable to nuisance switching (false-on) due to confusing air movement—near a supply grille, for example—with human motion.

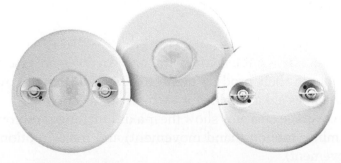

Figure 1-4. Ultrasonic (right), passive infrared (middle) and dual-technology (left) occupancy sensors. Photo courtesy of Watt Stopper/Legrand.

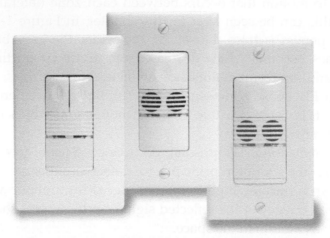

Figure 1-5. Ultrasonic sensors. Photo courtesy of Watt Stopper/Legrand.

Dual-technology sensors: Dual-technology sensors employ both PIR and ultrasonic technologies, activating the lights only when both technologies detect the presence of people.

This redundancy in method virtually eliminates the possibility of false-on and only requires either one of the two technologies to hold the lights on, significantly reducing the possibility of false-off. Appropriate applications include classrooms, conference rooms and other spaces where a higher degree of detection may be desirable.

Another type of dual-technology sensor combines PIR technology with acoustic detection to reduce the possibility of nuisance switching.

Coverage Area

Manufacturers publish range (ft.) and coverage area (sq.ft.) for their sensors in their product literature. Many different coverage sizes and shapes are available for each sensor technology.

The coverage area may show the maximum range and coverage area for minor motion (hand movement) and major motion (full-body movement).

The published pattern is often based on the maximum sensitivity setting for the sensor.

Figure 1-6. Dual-technology occupancy sensors. Photo courtesy of Watt Stopper/Legrand.

See Figure 1-7 for a sample coverage for a typical PIR wall switch sensor.

Mounting

Occupancy sensors can be mounted on ceilings, walls and within light fixtures and workstations. Below are common mounting configurations.

Ceiling: Appropriate for large areas that feature obstacles such as partitions, in addition to narrow spaces such as corridors and warehouse aisles (see Figures 1-8 and 1-9). Units can be networked for control of areas that are larger than what can be controlled by a single sensor. Typically 2-3 times higher installed cost than wall switch sensors, but can be very economical if controlling large zones.

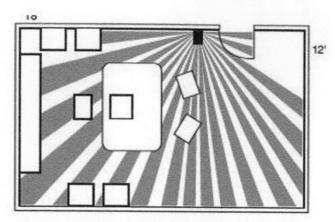

Figure 1-7. Coverage area for a typical PIR wall switch sensor. Source: Watt Stopper/ Legrand.

High wall and corner: Similarly appropriate for coverage of large areas that feature obstacles (see Figure 1-10).

Wall switch (wall-box): Appropriate for smaller, enclosed spaces such as private offices with clear line of sight between sensor and task area (see Figure 1-11). Relatively inexpensive and easy to install. For maximum savings, select sensors that provide positive disconnect from the power when the lights are off to avoid residual power use during off state.

Workstation: Appropriate for individual cubicles and workstations. The sensor is connected to a power strip for simultaneous control of lighting and plug-in loads such as computer monitors, task lights, radios and space heaters.

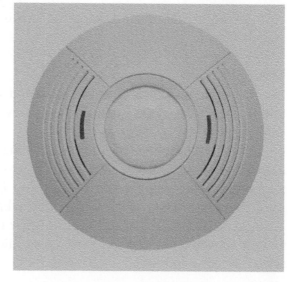

Figure 1-8. Ceiling-mounted occupancy sensor. Photo courtesy of Lutron Electronics.

Figure 1-9. Hi-bay occupancy sensor. Photo courtesy of Watt Stopper/ Legrand.

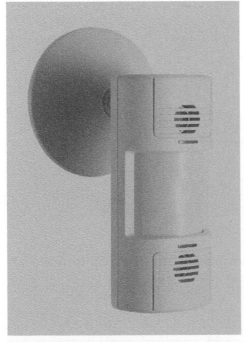

Figure 1-10. High wall and corner mount occupancy sensor. Photo courtesy of Lutron Electronics.

Light fixtures: Some occupancy sensors can be specified as an integral component of overhead light fixtures such as recessed baskets and direct/indirect fixtures (see Figures 1-12 and 1-13). This approach enables overhead sensors to be used without the fixture becoming an obstruction to the sensor, simplifies installation, and removes visual clutter from the ceiling plane.

Special Features

Depending on the manufacturer, a number of special features may be available for its products that can be used to optimize

Figure 1-11. Wall switch occupancy sensor. Photo courtesy of Lutron Electronics.

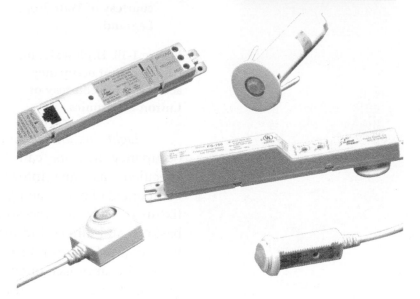

Figure 1-12. Fixture-mounted occupancy sensors. Photo courtesy of Watt Stopper/Legrand.

their application.

Manual-on operation: The most specified control option for occupancy sensors is automatic-on—the sensor turns on the lights automatically when a person enters the space. Some sensors, however, are available with a switch and require manual-on

Figure 1-13. Fixture-mounted occupancy sensors architecturally blend into lighting hardware without adding to visual clutter at the ceiling plane. Photo courtesy of Lightolier.

operation. This can increase energy savings because the occupant has the option to not turn on the lights because of available daylight or task lighting. However, occupants generally prefer auto-on operation.

Manual-off override: These sensors allow occupants to switch lights off manually using an integral switch. Sensors with this feature are typically used in spaces where the lights need to be switched off to view presentations.

Combination dimmer/occupancy sensor: Some wall-box sensors combine the functionality of an occupancy sensor and dimmer. The lights can be switched or dimmed based on occupancy. Dimming fluorescent and HID lighting requires a compatible dimmable ballast.

Masking labels: PIR sensors may be available with masking labels that allow the coverage area to be fine-tuned to prevent false-on triggering. For example, if a sensor's coverage monitors a private office but also an adjacent corridor, then the masking label can be used to obstruct the sensor's line of sight to the corridor.

Daylight switching: Some sensors can work with a light sensor to turn off the lights in response to sufficient ambient daylight and/or prevent the lights from reactivating as long as sufficient daylight is available. The setting is typically adjustable and can be overridden.

Isolated relay: Some power-packs and/or sensors contain a separate small low-voltage switch for control of and interfacing with additional loads such as HVAC, security and building automation systems. For example, people entering a building after hours trigger not only the required lighting, but also heat or air conditioning as well.

Network connectivity: Some sensors contain network communication capability that allows the sensor to connect directly to a digital network. This connectivity allows the sensor to operate other devices connected to the network and provide central monitoring capability.

Bi-level switching: Bi-level switching is encouraged by most energy codes (states that have adopted a version of IECC). Some power-packs include two separate relays for controlling two circuits simultaneously or independently (see Figure 1-14). This allows the sensor to integrate two manual switches for bi-level switching, which saves energy. For example, in a four-lamp troffer, one switch could control the ballast powering the outboard lamps, and the other switch could control the ballast powering the inboard lamps.

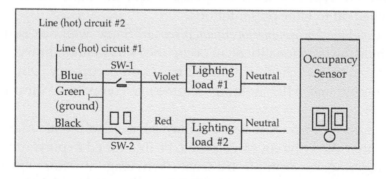

Figure 1-14. Occupancy sensors can be specified with a power pack that has two separate relays for controlling two circuits simultaneously or independently—enabling the integration of two manual switches for bi-level switching. Source: New Buildings Institute.

Evaluating Occupancy Sensors

Many occupancy sensor project failures can be traced to improper equipment selection and placement. Before specifying or purchasing a specific name-brand occupancy sensor, it may be beneficial to conduct a trial installation.

In a sample room, install the sensors temporarily. The sensors can be connected to the power supply but do not need to be connected to the lighting load.

Use the LED indicator on the sensor, which lights when the sensor detects motion. Move around the room and test the sensor's capabilities at different settings, varying the extent, speed and direction of your movement. Pay particular attention to whether the sensor can detect minor motion and if it is affected by false signals (such as activity in a corridor outside a controlled private office). If the sensor is ultrasonic, test its ability to detect motion behind obstacles.

Light loggers: Since energy savings delivered by occupancy sensors can be highly variable, developing an economic analysis often involves a number of assumptions. For greater confidence, some manufacturers offer devices that are temporarily installed in an existing facility for which occupancy sensors are being considered. These devices analyze how long the lights were left on while the monitored space is unoccupied. The data can then be downloaded into software for more accurate energy savings projections and analysis.

APPLICATION

A good lighting design requires a good control design.

Now that the basics of available occupancy sensor technologies are understood, the tools have been acquired to begin applying the sensors.

In this section, we will cover the seven primary steps involved in occupancy sensor design and application—including assessing the space characteristics, matching the right sensor to the application, layout and specification.

These steps are:

Step 1: Understand the application. Define the characteristics of the space and determine applicable energy code requirements.

Step 2: Choose the right sensor technology. Determine whether PIR, ultrasonic or dual-technology sensor(s) are best matched to the space characteristics.

Step 3: Select coverage pattern. Determine the range and coverage area for the sensor(s) based on the desired level of sensitivity.

Step 4: Select mounting configuration. Determine whether sensor(s) should be installed at the wall switch, wall/corner, ceiling or task.

Step 5: Layout. Locate the sensor(s) for optimal results.

Step 6: Specify the equipment. The technology, coverage area and mounting configuration have been determined. Now determine other features for the sensor(s) and the power-pack. Determine whether the sensor(s) must be integrated with other control devices.

Step 7: Installation and commissioning. Commission the sensor(s) by tuning their adjustable features.

Step 1: Understand the Application

The first step in sensor application is to understand the application characteristics. This involves understanding the application need, applicable energy codes and the space characteristics.

PIR sensors, for example, are most compatible with smaller, enclosed spaces (wall switch sensors); spaces where the sensor has a view of the activity in the space (ceiling and wall-mounted sensors); outdoor areas; and warehouse aisles.

Incompatible application characteristics include low motion levels by occupants; obstacles blocking the sensor view; and location within 6-8 ft. of HVAC air diffusers and other heat sources.

Ultrasonic sensors are most compatible with open spaces, spaces with obstacles, restrooms and spaces with hard surfaces. Incompatible application characteristics include high ceilings, high levels of vibration or airflow, and open spaces that require selective coverage (such as individual warehouse aisles).

Table 1-3 describes space characteristics that are relevant to application of PIR and ultrasonic sensors.

Table 1-3. Relationship of space characteristics and sensor technology.

Space Characteristics	PIR	Ultrasonic
Types of tasks performed in space.	Not as effective for low motion motion activity (dual-technology recommended for very low motion activity).	More effective for low motion activity (dual-technology recommended for very low motion activity).
Room/Space size and shape. Location of occupant tasks. Location of walls, windows and hanging objects. Location of shelves, book cases and large equipment.	Cannot see through obstacles such as corners. Limited sensitivity to minor (hand) movement at distances >15 ft.	Can see around corners. May need direct line of sight if soft surfaces such as fabric partition walls prevalent. Limited sensitivity to minor (hand) movement at distances >25 ft. Airflow from open windows or causing movement in hanging objects like drapes may cause false-on.
Partition height and location.	Cannot see through obstacles such as partition walls.	Ceiling-mounted sensor effective range declines proportionally to partition height.
Location of doors.	Sensor should activate lights as soon as person enters room. Sensor should not monitor area outside door to avoid nuisance switching. Door swing should not obstruct view.	Sensor should activate lights as soon as person enters room. Sensor should not monitor area outside door to avoid nuisance switching. Door swing should not obstruct view.
Ceiling height.	Sensitivity decreases proportionally to distance between person and sensor.	Less effective at ceiling height mounting locations > 14 ft.
Location of HVAC ducts and fans.	To avoid false-on, should not be located within 6-8 ft. from a HVAC air diffusers and other heat sources.	Sensor may experience false-on due to heavy airflow.
Potential sources of vibration.	Do not locate on potential sources of vibration.	Do not locate on potential sources of vibration.

Step 2: Choose the Right Sensor Technology

Determine whether PIR, ultrasonic or dual-technology sensors are best matched to the space characteristics.

Table 1-4 and Figure 1-15 provide guidance.

Table 1-4. General comparison of PIR and ultrasonic occupancy sensors.

Method	PIR	Ultrasonic
Coverage	Line of sight	Covers entire space (volumetric)
	Field of view can be adjusted by user	Field of view cannot be adjusted by user
Detects hand movement	Up to 15 ft.	Up to 25 ft.
Detects arm and upper torso movement	Up to 20 ft.	Up to 30 ft.
Detects full body movement	Up to 40 ft.	Up to 40 ft.
Coverage area	300-1000 sq.ft.	275-2000 sq.ft.
Highest sensitivity	Motion lateral to the sensor	Motion to and from the sensor
Mounting	Wall switch, wall, ceiling	Wall switch, wall, ceiling
Indoor/Outdoor use	Indoor, outdoor	Indoor

Step 3: Select Coverage Pattern

Manufacturers publish range (ft.) and coverage area (sq.ft.) for their sensors in their product literature. Many different coverage sizes and shapes are available for each sensor technology.

Coverage area, which depends on the level of sensitivity setting of the sensor, is a key application consideration. Improper coordination between required coverage area and required sensitivity is a leading cause of application problems.

Most manufacturers publish coverage areas based on the maximum sensitivity setting (this may not be stated, so ask). If based on the maximum sensitivity setting, the range and coverage area may need to be de-rated if sensitivity will be reduced from the maximum setting in your application.

The coverage area may show the maximum range and coverage area for minor motion, such as hand movement, and major motion, such as full-body movement.

It is critical to understand prior to selecting a sensor the full

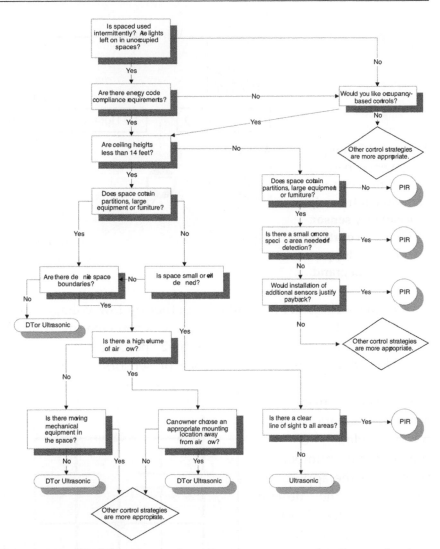

Figure 1-15. Decision map of occupancy sensor technology selection. Graphic courtesy of Watt Stopper/Legrand.

range of common tasks that will be performed in the space as well as environmental factors such as the location of furniture.

In a small space, one sensor may easily provide sufficient coverage. In a large space, however, it is recommended to partition the lighting load into zones, with each zone controlled by one

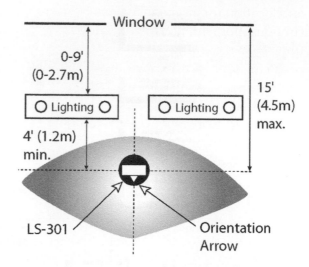

Figure 1-16. Sample coverage pattern for a ceiling-mounted ultrasonic occupancy sensor. Graphic courtesy of Watt Stopper/ Legrand.

sensor; the sensors are networked together by low-voltage wiring (see Figure 1-17). It is recommended that when creating zones, to ensure that sensor coverage overlaps by 20 percent.

Note that reducing the number of sensors to reduce cost can result in improper performance from the sensor installation. It is recommended to use the optimal number of sensors for the application to achieve the desired results. Manufacturers can provide application support including project layout and sensor location services.

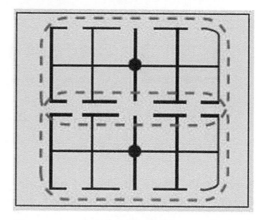

Figure 1-17. Networking occupancy sensors to achieve coverage of a large space. It is recommended to ensure that sensor coverage overlaps by 20 percent when creating zones. Source: Watt Stopper/Legrand.

Step 4: Select Mounting Configuration

Determining the mounting configuration

and location for a sensor is a critical design decision. The mounting configuration, in particular, affects the available coverage area. Improper mounting location is a leading cause of application problems with occupancy sensors. Earlier in this chapter, the most common mounting configurations were described: ceiling, high wall and corner, wall switch (wall box), workstation and integral with light fixtures.

Ceiling-mount sensors are best suited for open partitioned areas, small open offices, file rooms, copy rooms, conference rooms, restrooms, garages and similar spaces. The sensor provides 360° coverage; de-rate range by 50 percent if partitions >48 in. are in place.

Corner mount/wide view sensors are best suited for large office spaces and conference rooms. Mount these sensors high on the wall.

Wall switch sensors are best suited for private offices, copy rooms, residences, closets and similar spaces. These sensors are especially suitable for retrofits. They are not recommended for areas with obstructions that can block the sensor's view.

Narrow view sensors are best suited for hallways, corridors and aisles. These sensors work best if mounted on centers with range control.

High mount narrow view sensors are best suited for warehouse aisles. These sensors must be set back from the aisle so that they do not detect motion in cross aisles.

Table 1-5 provides additional guidance concerning occupancy sensor mounting.

Step 5: Layout

Locating the sensor is a critical design decision. Sensors should be located so that they have the least possibility of false-on/off switching, activate the lights as soon as the person enters the space, and have a permanent unobstructed line of sight to the task areas. Another aspect of location is orientation. For example, the receiving side of ultrasonic sensors should be positioned toward the area of greatest traffic in a space.

Table 1-5. Occupancy sensor mounting configurations. Adapted from 2001 *Advanced Lighting Guidelines*, New Buildings Institute.

Mounting Location	Sensor Technology	Angle of Coverage	Typical Effective Range*	Optimum Mounting Height
Ceiling	US	360°	500-2000 sq.ft.	8-12 ft.
Ceiling	PIR	360°	300-1000 sq.ft.	8-30+ ft.
Ceiling	DT	360°	300-2000 sq.ft.	8-12 ft.
Wall switch	US	180°	275-300 sq.ft.	40-48 in.
Wall Switch	PIR	170-180°	300-1000 sq.ft.	40-48 in.
Corner wide view	PIR/DT	110-120°	To 40 ft.	8-15 ft.
Corner narrow view	PIR	12°	To 130 ft.	8-15 ft.
Corridor	US	360°	To 100 ft.	8-14 ft.
High mount	PIR	12-120°	To 100 ft.	To 30 ft.
High mount corner	DT	110-120°	500-1000 ft.	8-12 ft.
High mount ceiling	DT	360°	500-1000 ft.	8-12 ft.

*Sensitivity to minor motion may be substantially less than noted above, depending on environmental factors.

PIR = passive infrared, US = ultrasonic, DT = dual-technology

In open offices that include partitions, ceiling-mounted sensors may be required. The maximum horizontal spacing between ceiling-mounted sensors may need to be de-rated based on the height of the partitions. If the partitions are higher than 48 in., for example, the range of ceiling-mounted sensors is effectively reduced by 50 percent. Consult the manufacturer on recommended spacing between sensors for partitioned areas.

Let's look at an application example. The drawing in Figure 1-18 portrays a large office area with partitioned cubicles. Primary tasks are performed at computers in individual workstations.

In this space, ceiling-mounted ultrasonic sensors are selected due to two requirements. First, the sensor must be able to "see" around obstacles. Second, the sensor must be able to detect minor motion, such as typing.

For reliable detection, sensors are located in zones that overlap by at least 20 percent when using the minor motion coverage (see Figure 1-18).

The application example in Figure 1-19 portrays a public restroom with six partitioned stalls. The sensor must be able to see around obstacles.

A ceiling-mounted ultrasonic sensor can be located about 2 ft. out from the stall door to cover the entire space.

The application example in Figure 1-20 presents a series of warehouse aisles with high ceilings. The lighting in each aisle is configured as a separate control zone so that the sensors only trigger the lights in the aisle that is occupied.

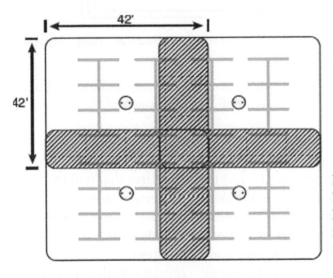

Figure 1-18. Layout example, large office space. Source: Watt Stopper/Legrand.

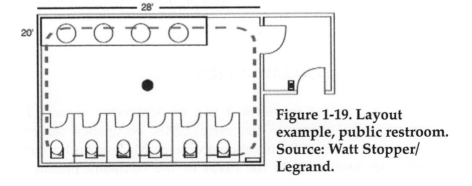

Figure 1-19. Layout example, public restroom. Source: Watt Stopper/ Legrand.

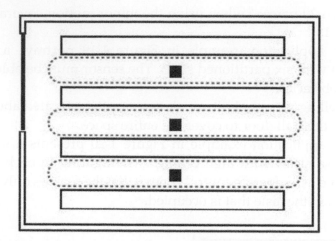

Figure 1-20. Layout example, warehouse. Source: Watt Stopper/ Legrand.

In this space, a ceiling-mounted PIR sensor can be installed in each aisle and its coverage area defined for a narrow view.

Step 6: Specify the Sensor

The occupancy sensor can now be specified. Be sure to examine potential special features and select those features that will result in optimal value based on the combination of performance and cost.

It's generally desirable to select a manufacturer that offers strong application support and whose products have a strong track record of reliability, performance and innovation.

The power pack is a significant component that is part of the specification. Table 1-6 provides a description of common specification variables.

COMMISSIONING OCCUPANCY SENSORS

During the first widespread use of occupancy sensors in commercial applications in the early 1990s, occupancy sensors experienced a number of application problems. Many in the lighting industry are familiar with the myth of the company president

Table 1-6. Specifying the power pack.

Power output	The consumption of devices connected to the power-pack should not exceed its output rating.
Voltage	Some power-packs operate on only one voltage while others are multi-voltage.
Relay type	Basic model is an A relay (normally closed) that is isolated so it can switch a load differing from the line voltage feeding the power-pack; other options include two or more isolated relays in same unit for simultaneous or independent control of two or more loads, and smaller low-voltage relays that can be used to signal other systems such as security and HVAC.
Relay ratings	Most power-packs can switch a 20A load. Some do not. Be sure the load does not exceed the power-pack's load rating. For extended life, it is recommended to use relays with zero-crossing circuitry.
Type of load to be controlled	Power-pack ratings may differ based on the type of lighting being controlled. Be sure the load does not exceed the power-pack's ratings for the given load type.
Mounting	Power-packs are typically installed in the plenum, mounted on a junction box. Be sure the power-pack and control wiring is rated for plenum installation.
Signaling from multiple devices	Smart power-packs feature inputs for control of the load by multiple control devices.

sitting in a restroom stall when the lights went out. Since the early 1990s, occupancy sensor technology and products have improved dramatically. Today, occupancy sensors provide reliable operation without disrupting business activity if the sensors are properly specified, installed and calibrated.

Today's application problems usually occur due to improper sensor selection, mounting or calibration. For this reason, following proper installation according to manufacturer wiring diagrams and applicable codes, commissioning the occupancy sensor system is essential to achieving optimal results for the application.

Commissioning entails systematically testing all controls in the building to ensure that they provide specified performance and interact properly as a system, so as to satisfy the design intent and owner's needs. It is vital to ensure proper operation, user acceptance, and energy savings potential in new construction as well as renovation and retrofit projects.

The first step in commissioning an occupancy sensor installation is to verify proper placement and, if applicable, orientation of the sensors, so they match the specifications and construction drawings.

The second step in commissioning an occupancy sensor installation is to calibrate the sensor by adjusting its sensitivity and time delay settings. Calibration is a standard tool used in commissioning. Calibration means adjusting sensors from factory default settings to gain desired performance based on the actual conditions of the application. It should be coordinated with furniture placement, as occasionally furniture or equipment may be moved or relocated, which can affect sensor placement and/or orientation.

In this section, we will cover how to commission and calibrate occupancy sensor installations and maintain these devices for successful long-term use.

Calibration: Motion Sensitivity

The occupancy sensor's sensitivity level indicates how much movement will cause the sensor to activate the lights. If you have too high a level of sensitivity, you can increase the possibility of false-on triggering. If you have too low a level, you can increase the possibility of false-off triggering. Sensitivity is also related to coverage; changing the sensitivity can result in changes to the coverage pattern.

Occupancy sensors are shipped with a factory setting for sensitivity, which can be adjusted in the field to fit the application need. The sensitivity adjustment allows the contractor to tune the sensor so that it responds appropriately to primary tasks in the space at the designed distance, while addressing possible sources of nuisance switching such as airflow. Setting the sensor at the most appropriate sensitivity setting will minimize the possibility of nuisance switching.

Calibration: Time Delay

The occupancy sensor's time delay indicates how much time it will take to shut off the lights after the last motion has been detected. Longer time delays avoid continual on/off cycles of the lighting as people may go in and out of the space frequently.

Time delays also help to overcome brief periods of time when activity levels are low, as the sensor may not detect the occupant and therefore inadvertently switch the lights off.

Typical factory settings can be anywhere up to 20 minutes, which can be adjusted in the field. The shorter the time delay, the greater the energy savings but also the shorter the lamp life, due to more frequent switching, which results in more wear and tear on the lamps. In many applications, time delays of no less than 15 minutes are recommended.

Self-calibrating Sensors

Some occupancy sensors are self-calibrating and therefore require little or no adjustment of their sensitivity and time delay settings (see Figure 1-21).

This "install and forget" sensor contains a microcontroller that continually monitors the space to identify usage patterns. Using this information, the sensor automatically adjusts its sensitivity and time delay settings for optimal performance based on these usage patterns. Over time, environmental conditions may change, but this type of sensor will continually adjust its operation to provide consistent service.

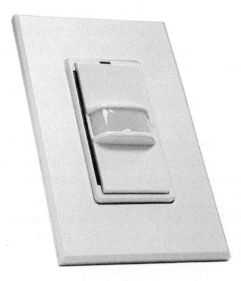

Figure 1-21. Self-calibrating occupancy sensor. Photo courtesy of Lightolier Controls.

It is important to understand, however, that a self-calibrating sensor may inadvertently switch off lighting during the early stages of monitoring usage patterns. Re-triggering by detecting motion immediately after switching off the lights causes the sensors to adjust time delay, sensitivity or both.

Commissioning

The last step in commissioning an occupancy sensor installation is to test its performance; several simple tests can be performed.

Entry test: If the sensor is used to automatically turn on the lighting, then it should detect a person entering the space within one to two steps of entry into the space. The sensor should only switch on lighting once a person has committed to entering the space and not turn on the lights when a person passes outside the room with the door open.

Hand motion test: The sensor should turn the lights on if you wave your hand in different directions a foot away from the sensor.

Perimeter test: Walk and wave your hand in different places around the room to try to find spots where the sensor is least effective in detecting motion.

After Commissioning

After commissioning is completed, tell the users about the intent and functionality of the controls, especially the overrides. This is critical because if users do not understand the controls, they will complain and attempt to override or bypass them. Give all documentation and instructions to the owner's maintenance personnel so that they can maintain and re-tune the system as needed.

In the future, occupancy sensors should be tested periodically for sensitivity and time delay settings to ensure design intent continues to be satisfied as environmental conditions change in the space over time.

It is recommended that maintenance personnel should inspect all lighting controls for proper operation at least once per year as part of their lighting maintenance procedures.

Design, Application and Commissioning Issues

Compatibility with electronic ballasts: When most lighting loads are switched on, they can draw a much higher level steady-state current for the first couple of line power cycles—a fraction of a second. In some electronic ballasts, particularly ones using front-end active filters to reduce harmonic distortion, this "in-rush current" may cause the contacts in lighting relays to pit and eventually fuse.

Older occupancy sensors were never designed with this in mind and could become damaged by in-rush current. More recent generations of occupancy sensors mitigate the effects of in-rush current using zero-crossing circuitry. The National Electrical Manufacturers Association (NEMA) has published Standard 410 for testing devices against the "worst case" in-rush current. This standard will provide the testing basis for the UL listing of these devices in the future. Until then, it is recommended that all sensors be required in specifications to employ zero-crossing circuitry, or get written confirmation from the manufacturer that the device has been tested for the in-rush current specified in NEMA Standard 410.

Energy savings versus lamp life: If the time delay setting for an occupancy sensor is shorter, energy savings will be maximized because the lights will be shut off for an overall longer period of time. In 1997, researchers (Maniccia et al 2000) studied energy savings potential for occupancy sensors in multiple space types and buildings and tracked energy savings versus time delay.

The research indicates that the shorter the time delay, the greater the energy savings. However, the shorter the time delay, the more starts that are likely to occur, which can reduce lamp life.

Shorter time delays can increase the amount of switching of the lamps, which reduces their average rated life, as most wear and tear on the lamp is caused during startup. This does not necessarily mean lamp costs will be higher. The lamps may provide fewer burn-hours, but may not provide a shorter calendar life because they will be shut off for longer periods of time than if they were continuously operated. If the lamps must be replaced more frequently, the premium is typically less than the energy savings.

Research has shown that sensible lamp life can be achieved with time delays of 15-30 minutes. Nonetheless, this is a factor that should be addressed in the economic analysis. For example, if the energy rate is $0.08/kWh and the cost of replacing one lamp is $6.60 in labor and materials, then the break-even point is a time delay of about 10 minutes.

To optimize lamp life, it is recommended that a minimum 15-minute delay setting should be used for occupancy sensors, that only ballasts that meet ANSI requirements for lamp ignition should be used, and that programmed-start ballasts should be used in applications with a high number of switching cycles per day and where lamp life is a primary concern. Programmed-start ballasts have gained in popularity due to their utility in applications where the lamps are frequently switched, such as occupancy sensor installations. Programmed-start ballasts are rapid-start ballasts that start fluorescent lamps with less wear and tear on the lamp cathodes, which increases lamp life. As a result, programmed-start ballasts can provide up to 100,000 starts, ideal for occupancy sensor installations.

Backup lighting: It is recommended that emergency lighting not be controlled by an occupancy sensor. Emergency backup lighting will only come on during normal power interruption and is not influenced by a sensor failure.

Ultrasonic sensors and hearing aids: This section addresses certain marketing and sales claims made in the industry about ultrasonic sensors by at least one U.S. manufacturer that appear to be false.

Hearing Aids: There have been claims that ultrasonic sensors interfere with the operation of hearing aids. NEMA member companies report less than five incidents per year of any type of hearing aid interaction; the possibility for interaction is remote, according to manufacturers. Hearing aid manufacturers also report that customer support calls about hearing aid interference with ultrasonic sensors are very limited. Ultrasonic sensors that operate at 32 kHz and above do not interfere with hearing aids.

There has been a claim that hearing aids detect the emission of ultrasonic sensors and stress the hearing aid device's battery, causing it to fail prematurely. Another claim is that when the hearing aid detects the emission of ultrasonic sensors, it triggers the automatic gain control feature of the aid, lowering its detection so that sounds in the normal audible range are not easily heard.

Independent research by NEMA has demonstrated that these claims are not true for modern hearing aids.

The NEMA-sponsored study, conducted by David F. Henry, PhD and Barak Dar, "Effects of Ultrasonic Sensors on Hearing Aids" (February 2006), concluded:

> **"After assessing 23 hearing aids representing the digital products of all major hearing aid manufacturers, just two hearing instruments were severely affected by ultrasonic occupancy sensor devices. One device, the Impact DSR675 manufactured by AVR Sonovation, has not been in production for over four years. Newer hearing instruments from this manufacturer have exhibited no interference when exposed to the ultrasonic occupancy sensor signals. The second device, the Bravo, manufactured by Widex, is still in production. However, the manufacturer reports that they have developed a modification that can be made to the instrument that greatly reduces the susceptibility of the instrument to ultrasonic occupancy sensor signals."**

When interaction occurs, it is typically the result of a hearing aid that has fallen into disrepair or has drifted out of its spec, as determined by NEMA findings which have been reinforced by independent testing. Hearing aids should be periodically re-calibrated to within specifications to avoid interference problems. A hearing aid that exhibits interference problems should be returned to the patient's audiologist or clinician for adjustment. In all cases, the problem can be resolved quickly and easily.

Ultrasonic operation, however, may result in minor noise in the hearing aid, although this does not affect hearing aid performance. Henry and Dar add: "Several [hearing aid] instruments from other manufacturers emit minor noises when exposed to the ultrasonic occupancy sensor devices, but not to the extent of rendering them inoperable, or affecting their performance." The potential for minor noise is typical of electrical devices. For example, cell phones, microwave ovens, electric motors and many other electrical devices can generate audible noise in hearing aids. Compared to many other devices, ultrasonic sensors are a minor contributor, say manufacturers.

The frequency and decibel levels used with ultrasonic sensors fall well within the safety guidelines of the U.S. Food & Drug Administration, the World Health Organization, and various governments of industrialized nations around the world.

Troubleshooting Nuisance Switching

Troubleshooting is much like detective work, in which one has a problem and then determines a list of likely suspects, narrowing the list through a process of elimination until the cause is determined. In this section, possible causes and solutions are detailed for a number of potential problems related to nuisance switching in occupancy sensors.

Should nuisance switching occur, consult the false-on/off switching troubleshooting guide in Table 1-7.

Occupancy Sensor Commissioning Study

A 2007 study conducted by ZING Communications, Inc., co-sponsored by the Lighting Controls Association and Watt Stopper/Legrand, suggests that electrical contractors routinely calibrate motion sensitivity and time delay settings in occupancy sensor installations; recommend occupancy sensors in a majority of lighting retrofit projects; select time delay settings that on average support optimal energy savings and lamp life; and are satisfied with occupancy sensor performance, ease of installation and commissioning, and customer/occupant acceptance.

However, the survey results also suggest that contractors nonetheless report a high rate of callbacks compared to previous research concerning other lighting technologies. The typical solution is to return to the job to adjust the time delay and sensitivity settings, according to the results. This situation suggests a strong potential for the utility of self-calibrating occupancy sensors. The ability of

Table 1-7. Guide to troubleshooting false-on/off switching in occupancy sensors. Source: Lighting Controls Association, 2006.

Problem	Possible Causes	Prescription
False-on (PIR sensors)	Sensor is detecting heat from an artificial source	Sensor should not be located within 6-8 ft. of HVAC outlet, heating blower or other heat source; applying masking on the lens may reduce or eliminate the problem
	Sensor is detecting motion in an area adjacent to the control zone	Re-orient sensor if high wall- or corner-mounted, use masking labels to restrict coverage area, or move the sensor; bear in mind that some overlap detection may be desirable for people sitting in fringe areas
	Sensor is detecting motion in an adjacent warehouse aisle	Use narrow-view sensor so that it "sees" only the aisle in its control zone Use masking labels to customize the field of view
	Re-commission the installation.	

Problem	Possible Causes	Prescription
False-on (ultrasonic sensors)	Heavy airflow from HVAC outlet, open window or other source	Sensor should not be placed within 4-6 feet of an air supply register
	Vibration	Ensure the sensor is mounted on a vibration-free, stable surface
	Object is moving in the space	Ensure drapes and other hanging objects such as mobiles or flags are not moving due to airflow
	Sensor is detecting motion in an area adjacent to the control zone	Re-orient or move the sensor; bear in mind that some overlap may be desirable for people sitting in fringe areas
		Ultrasonic sensors are suitable for restricted coverage in closed areas, such as corridors—use a PIR sensor for open areas requiring restricted coverage, such as warehouse aisles
	Re-commission the installation.	

Problem	Possible Causes	Prescription
	Very low levels of occupant motion	Verify that time delays are set to no less than 15 minutes or increase delay up to 30 minutes. Increase sensor sensitivity
		Use dual-technology sensor for greater sensitivity/reliability
		Check the range and coverage area of the sensor; note the sensor's range and coverage area is affected by the sensitivity setting

False-off (PIR sensors)	Occupant is too far from sensor	Check the sensitivity level and adjust if too low and adjustment is acceptable
		Place the sensor as close as possible to the main activity
		The use of ceiling sensors typically avoids this problem
		Move the task, object or replace with an ultrasonic sensor, which does not require a direct line of sight
		Ensure sensor is not installed where a door swing, filing cabinet or bookcase will obstruct view
	Re-commission the installation.	
Problem	*Possible Causes*	*Prescription*
False-off (ultrasonic sensors)	Very low levels of occupant motion	Verify time delays are set to no less than 15 minutes, or increase delay to 30 minutes
		Ultrasonic sensors are more sensitive than PIR sensors, but very low levels of occupant motion may still result in false-off; consider replacing with a dual-technology sensor
	Ceiling-mounted sensor mounted too high	Sensor should not be mounted higher >14 ft.
	Occupant is too far from the sensor	Check the range and coverage area of the sensor; note the sensor's range and coverage area is affected by the sensitivity setting
		Check the sensitivity level and adjust if too low and adjustment is acceptable
	High partitions, bookshelves or equipment	High partition walls can restrict range of sensors; the coverage area must be de-rated and the sensors spaced closer together as needed
		Place the sensor as close to the main activity as possible
		If sensors are networked to control large area, at least 20 percent overlap in coverage patterns is recommended
	Soft surfaces such as fabric-walled partitions are resulting in poor reflection of ultrasonic waves	Provide a clear line of sight between the sensor and the occupant
		Consider 30-minute time delay
		Consider additional sensors
	Re-commission the installation.	

self-calibrating sensors to reduce callbacks has not been studied, however, creating a demand for further research.

Research Questions

A random sampling of typical factory settings for occupancy sensors indicates typical settings of 10 or 30 minutes.

Manufacturers recommend that contractors commission the sensors upon installation, and reset the time delay to a setting appropriate for the application.

This raised the question: Do contractors typically commission sensors, and reset the time delay, and if so, what time delay settings are typically selected?

As the installed base of occupancy sensors rapidly grows, this question becomes highly relevant to the larger question of how much energy is being saved.

Research Study Addresses Time Delay/Commissioning

In May 2007, ZING Communications, Inc. invited 450 electrical contractors to take an online survey asking them questions about occupancy sensor commissioning and performance. This population was pulled from the subscriber list of LightNOW and lightingCONTROL, two industry e-newsletters. Thirty qualifying respondents completed the survey, a 6.9 percent response rate. To qualify, respondents indicated they install lighting products in commercial projects in North America, and have installed occupancy sensors in the past. The results reflect, with a high degree of statistical accuracy, the views of the studied population. Note, however, this population has an inherent bias because it is constituted of subscribers to lighting newsletters, which brings up the possibility that the population is better educated about controls than other contractors. Keeping this bias in mind, the results are suggestive of the electrical industry.

Respondents were asked if they commission occupancy sensors after installation. For the purposes of this survey, commissioning is defined as calibrating the motion sensitivity and time delay settings of the sensor.

Nine out of 10 respondents (90.3 percent) report that they commission occupancy sensors after installation.

Respondents were then asked what time delay they typically set. Responses varied. About one-third of respondents (32.3 percent) report a 10-minute time delay as the typical setting, while about one-sixth of respondents (16.1 percent) respectively report selecting 5 minutes, 15 minutes and 20 minutes. More than one in 10 respondents (12.9 percent) select 30 minutes. Less than one in 10 (6.5 percent) typically don't set the time delay, but retain the factory default setting. Discounting those that retain the factory default setting, the weighted average time delay for occupancy sensors installed by respondents is 13.5 minutes, suggesting that good energy savings are being realized.

Despite commissioning, the average respondent reports being called back on about one in five projects (18.7 percent) to make further adjustments or commissioning of occupancy sensors. This is about twice the callback rate that electrical contractor respondents reported in a separate 2005 survey about fluorescent dimming systems (9 percent), and more than twice the callback rate that electrical contractor respondents reported in another 2004 survey about all industrial/commercial lighting projects (7 percent).

The survey suggests that return visits to a site is a major problem with some contractors. Three responses—100 percent, 85 percent and 75 percent—increased the average callback rate (if these responses were removed, the callback would drop to 11 percent).

The survey further suggests that the time-delay settings of the sensors may be a factor. Respondents who report setting the time delay of the installed sensor to 10 minutes or less also report a much high rate of callbacks (22 percent) than those setting a time delay of 15 minutes or longer (14 percent).

When called back to resolve a problem with an occupancy sensor, nine out of 10 respondents (about 90 percent) say they adjust the time delay and/or sensitivity settings to correct the problem. About 10 percent of respondents say they typically replace the sensor. None of the respondents report moving the sensor as a solution. (Also note that in the below graphic, the "other" response was a respondent reporting they had never had a callback; the other numbers were adjusted accordingly to about 90 percent/about 10 percent. Additionally, no "other" responses involved correction of miswiring or other installation errors.)

Satisfaction with Sensors

Electrical contractors responding to the survey indicate they are more than somewhat satisfied, on a 1-7 scale, with 1 being "not satisfied," 4 being "somewhat satisfied," and 7 being "very satisfied," with these attributes of occupancy sensors:

- Customer/Occupant satisfaction (5.66)
- Energy savings (5.59)
- Ease of installation (5.38)
- Reliability (5.31)
- Ease of application (selecting the right sensor, locating it, time delay, sensitivity) (5.14)
- Ease of commissioning (4.83)

In addition, all respondents report that their firms make recommendations on lighting equipment choices in lighting upgrade projects in existing buildings. The average respondent says his/her firm recommends occupancy sensors in 60.7 percent of the firm's lighting upgrade projects in existing buildings.

Final Conclusions Suggested by Research

- The average electrical contractor is more than "somewhat satisfied" with occupancy sensor performance and ease of installation and commissioning.

- The average electrical contractor regards his or her customers and associated occupants as more than "somewhat satisfied" with occupancy sensor performance.

- The average electrical contractor is most satisfied with

customer/occupant satisfaction and energy savings, and least satisfied with ease of application and ease of commissioning.

• The average electrical contractor recommends occupancy sensors in a majority of lighting upgrade projects in existing buildings.

• Occupancy sensors are typically commissioned in terms of calibration of motion sensitivity and time delay settings.

• Time delays are, on average, set within an optimal range to maximize both energy savings and lamp life.

• The average electrical contractor experiences a high callback rate for further adjustments to the sensor, which may argue for longer time delay settings (about one third of respondents, for example, set the time delay for 10 minutes); a minimum of 15 minutes is recommended.

• During callbacks, the contractor most often further adjusts the motion sensitivity and time delay settings to correct the problem.

• Self-calibrating sensors have been designed to reduce both commissioning requirements and the callback rate; the utility of these devices need to be validated in further study.

Chapter 2

Switching Controls

Switching controls activate and deactivate connected lighting loads. These on-off controls can be as simple as a manually operated light switch, a timer, panel-boards containing contactors, or a relay mounted in a light fixture.

While manual switches are common in most application types, the trend is toward intelligent automated local switching and remote automated switching—resulting in superior control precision, energy management and energy information. Savings of 5-15 percent for scheduling controls and 10-50 percent for occupancy-based controls (covered in Chapter 1) have been demonstrated.

While energy savings using on-off switching controls (with the exception of occupancy sensors) are often less than dimming controls due to longer on periods and/or time scheduling, switching systems are generally easier to commission and can be installed at a lower initial cost.

Switching controls can be as basic as a standard AC wall switch and as sophisticated as a lighting control network that performs multiple energy-saving strategies and can combine switching with dimming. Typical strategies and savings are summarized in Table 2-1. Since energy-saving switching strategies often have overlapping device requirements, significant synergies and economies can be achieved.

With the exception of manual bi-level switching, most installed energy-saving lighting control strategies are driven by automatic switching of loads. Automation of the on/off function is typically based on one of two inputs:

- Time—activating and deactivating lighting loads based on a schedule; or

- Threshold event—activating and deactivating lighting loads based on either daylight's contribution to the light level in the space, or the detected presence or absence of people (occupancy sensing, covered in Chapter 1).

Most automatic controls are microprocessor-based devices that offer superior intelligence and capabilities compared to traditional electromechanical devices.

Time scheduling: When scheduling is applied, lighting in given areas is turned on, shut off or adjusted to a predetermined schedule. In some cases, the systems control may be vested in a different device. For example, a given system may be under the direction of daylight harvesting controls from 9:00 AM through 4:00 PM, and under the direction of demand management controls from 11:00 AM to noon and 2:00 PM to 4:00 PM.

Scheduling is a time-based function and as a consequence it is most suited for facilities or spaces where the occupancy pattern is predictable and certain things happen at certain times. Because "off-normal" conditions inevitably arise, local overrides are usually provided.

Threshold-based control: Threshold-based control switches loads primarily based on two types of inputs:

• Threshold based on occupancy: A sensor detects the presence or absence of people in the space and

Figure 2-1. Intelligent lighting control panels enable scheduled switching or dimming to save energy based on time of day. Photo courtesy of Square D.

switches the load based on the occupancy status.

* Threshold based on light level: A sensor monitors light level and switches or dims the load based on the light level crossing a predetermined threshold.

Occupancy sensors, which control loads based on the occupancy threshold, were covered in Chapter 1. Switching based on light level is used in spaces where enough daylight enters the space during times of the day to become a primary or contributing source of ambient illumination. Using a strategy called daylight harvesting, energy savings can be gained by turning the electric lighting off or adjusting its output if sufficient daylight is present. Daylight harvesting is covered in detail in Chapter 8.

In this chapter, you will learn about the types of switching controls—from wall switches to integrated systems—in addition to wiring methods, integration with building automation systems, and how these controls are used to support energy-saving strategies.

SWITCHES

This section describes switches commonly used in commercial applications.

Power Switches: Lighting Contactors (and Relays)

Power switch devices permit manual or automatic control of lighting loads. Available types include lighting contactors— including relays, and remotely operated circuit breakers.

Lighting contactors: Three types of lighting contactors in common use are:

* Feeder-disconnect type (2-, 3- and 4-pole contactors rated from 20A to 600A to control large blocks of load);

* Multi-pole contactors with as many as 12 poles (rated 20-30A) for multi-branch circuit control; and

Table 2-1. Typical switching strategies.

Strategy	How It Works	Typical Energy Savings
Occupancy sensors	Turn the lights on or off automatically based on whether space is occupied	35-45 percent (PG&E); higher savings can be achieved in private offices and smaller spaces
Scheduled automatic shut-off at end of workday via switching panels, time clocks or a building automation system	Turn selected lights on or off automatically based on a schedule when the space is predictably unoccupied; usually features local override control	5-10 percent (California Energy Commission); typically higher if lights left operating 24/7 but facility occupied only for portion of day
Scheduled automatic shut-off of select loads (bi-level switching) during peak demand periods via control panels, time clocks or a building automation system	Turn off one or two lamps in each fixture or checkerboard pattern of fixtures automatically for load shedding during peak demand periods	5-15 percent (California Energy Commission); typically higher if lights left operating 24/7 but facility occupied only for portion of day
Bi-level or multi-level switching using manual wall switches controlling lighting system layered as two or more circuits	Turn selected circuit on and off manually to achieve 100%/50%, or 100%/66%/33% light level and off	10-15 percent (NEMA)
Multilevel switching using photosensor and low-voltage relay	Turn the lights off automatically based on available ambient daylight	

- Single-pole relays rated 20A with low-voltage control for individual branch circuit and light fixture control.

Relays can be specified with contacts normally open or normally closed or with contacts latching or mechanically held. Relays with normally closed contacts ensure continuation of electrical contact during a primary power failure. Relays used on emergency circuits must be UL-listed specifically for such use, however.

Contactors are used with many forms of automatic controls, as through integration with solid-state lighting control modules that operate as a function of a photosensor or occupancy sensor input or with microprocessor-based energy monitoring and control systems.

Remotely operated circuit breakers: Like contactors, remotely operated circuit breakers permit both manual and automatic control of branch circuits. They can be switched independently, providing individual branch circuit-level control, or grouped and switched together to control multiple branch circuits. They are commonly available in 1-, 2- and 3-pole versions with current ratings ranging from 15A to 60A.

Wall Switches

The basic local lighting control is a wall switch (AC snap switch), rendered in Figure 2-2. This switch can handle a full 20A branch circuit lighting load, such as 24 four-lamp fluorescent fixtures at 277V (after de-rating per electrical code). This basic control is manually operated and, with only two choices of light levels (100 percent and 0 percent), not very flexible.

Key-activated Switches

Key-activated switches are wall switches that turn lighting on and off using a key. They are installed to prevent unauthorized or accidental use of certain lighting circuits. They are particularly useful for high-intensity discharge (HID) light sources that must cool down after operation before they can be re-started. See Figure 2-3.

Bi-level Switching

Wall switches can be applied to develop a flexible lighting control scheme using a strategy called bi-level switching. Figure 2-4 shows a section of an office, lighted by eight 4-lamp fixtures. Two wall switches can be used, one to control all outboard lamps (A and D) while the other controls the inboard lamps (B and C). Alternately, one switch could be used to control all even-numbered fixtures and another for the odd-numbered, or to control fixtures

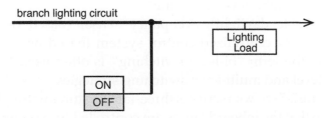

Figure 2-2. Simplified schematic for a basic wall switch.

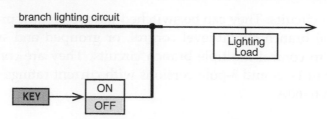

Figure 2-3. Simplified schematic for a key-activated wall switch.

Figure 2-4. Bi-level switching.

1/2 and 5/6 separately from the other four.

Any of these techniques permits a 50 percent reduction in light output, with the best selection being that which closely matches occupancy patterns in the space.

By using four controls, even more variations are possible.

While energy savings can be less than other strategies due to reliance on human initiative, bi-level switching is a simple, durable switching strategy.

Another strategy is multi-level switching, which can be initiated using a manual switch or automatically based on a control signal from a photosensor (based on detected light level resulting

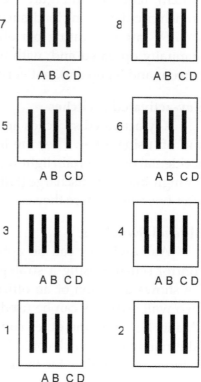

from available daylight), occupancy sensor (based on detected occupancy), or centralized control system (based on a schedule). Note that the term "bi-level switching" is often used to describe both bi-level and multi-level switching strategies.

Figure 2-5 shows a series of three-lamp fixtures with split-ballast wiring so that the inboard lamps are controlled on a separate circuit than the outboard lamps. This allows four light levels—100 percent (all lamps lit), 66 percent (2 lamps in each fixture lit), 33 percent (1 lamp in each fixture lit), and 0 percent (all lamps extinguished). The Figure shows two lamps being lit in each fixture, providing 66 percent light output. Multi-level switching provides greater flexibility than bi-level switching and poses a less abrupt change in light level when automatic control is used.

Figure 2-5. Multi-level switching.

Another option for controlling all lamps and fixtures is to use step-dimming or light level switching ballasts, which provide a uniform change in illumination in the space. For example, a light level switching ballast incorporates two hot power leads for control with two standard switches or relays; switching one lead on provides 50% power while having both switches on provides 100% power. Alternatively, if only one switch is available or desired, the light level switching ballast can provide 100% power when the switch is first turned on and 50% after toggling down and back up.

According to the New Buildings Institute, potential energy savings of 15 percent can be achieved with multi-level switching in classrooms and 30 percent can be achieved in private offices. A second study by ADM Associates found lower savings:

In May 2002, "Lighting Controls Effectiveness Assessment: Final Report on Bi-Level Lighting Study" was published by the California Public Utilities Commission (CPUC), prepared by ADM Associates for Heschong Mahone Group, project managers for the Southern California Edison Company on behalf of the CPUC.

This is one of only a few field studies that have actually examined the use and utility of bi-level switching as a means to reduce energy costs. Two specific goals of the study were:

- Study how occupants used manual bi-level switching controls, including behaviors that reduced savings potential; and

- Estimate energy and demand savings.

The researchers measured data for bi-level switching applications in 256 open and private office, retail and classroom spaces in 79 buildings. The fixtures contained three lamps that were switched in a multi-level switching scheme, providing four lighting states: all lamps off, 1/3 lamps operating, 2/3 lamps operating and all lamps on.

Table 2-2 shows the breakdown of use of different bi-level switching conditions (high-wattage or 2/3 lamps switch only on, low-wattage or 1/3 lamps switch only on, or both switches on or off).

Table 2-2. Use of bi-level switching conditions at 3PM on weekdays by space type. Source: ADM Associates

Type of Space	Both Switches Off	Both Switches On	High-Wattage Switch Only On	Low-Wattage Switch Only On
Open office	10.4%	65.8%	14.9%	8.9%
Private office	30.3%	47.9%	14.5%	7.3%
Retail	4.1%	78.9%	12.0%	5.0%
Classroom	57.6%	34.4%	5.3%	2.8%

ADM Associates discovered that private offices demonstrated the highest level of energy savings derived from using bi-level switching at 21.6 percent (with bi-level energy savings defined as occurring at 1/3 or 2/3 power). Open offices came in second at 16 percent, followed by retail at 14.8 percent and classrooms at 8.3 percent.

One of the factors of bi-level switching use that was studied was daylight contribution. Use of bi-level switching and subsequent energy savings in open offices and retail spaces showed a positive correlation with daylight availability. Private offices did not show a positive correlation. Classrooms did, but demonstrated the opposite of researcher expectations: Classrooms with the lowest amount of daylight also had the lowest level of use of lighting.

In the end, the study demonstrated that manual bi-level switching results in energy savings, which could be increased with occupant education, and with the limitations on the use of only one switch offset by the simplicity and economy of the approach.

Timer Switches

Another energy saving strategy is the timer switch (see Figure 2-6). These devices turn off the lights in a single load switch leg after a preset period of time once the lights have been switched on. These switches may be programmable electronic switches or spring-loaded, mechanical, twist-timer switches. The shutoff setting is determined by the user or the contractor, depending on the technology.

For example, a user enters a room, activates the lights and the timer, and when the timer expires, the lights shut off. When the lights are about to shut off, a warning signal may be emitted. This makes the timer switch both an occupancy- and time-based strategy to save energy that can be available for less than one-third of the cost of an occupancy sensor.

However, timer switches typically save less energy than occupancy sensors, and may experience nuisance switching, as the lights will shut off at the end of the period, unless the user restarts the timer. For this reason, timer switches are typically used in store rooms, mechanical and electrical rooms, supply closets and janitorial spaces.

Time Clocks

Time clocks, also called time switches, are used for simple time-based on/off control in small commercial buildings, common-area lighting in apartment buildings, and outdoor billboard and parking lot lighting. One of the most common applications is control of a single outdoor lighting load on a schedule—on at dusk, off at dawn. Used indoors, however, it could activate lighting several times each day—e.g., on at 8:00 AM, off it at noon, on at 1:00 PM, and off again at 6:00 PM.

A time clock may be a stand-alone control or may be integrated as a scheduling feature of lighting control panels, which are described later in this chapter.

Figure 2-6. Timer switch. Photo courtesy of Watt Stopper/Legrand.

Stand-alone time clocks may be electronic or electromechanical. Electromechanical time clocks may be designated as 24-hour or 7-day devices, with the 7-day type being more commonly used. If the lighting schedule does not change day to day, then the 24-hour model can be used. If the lighting schedule changes during the week—due to weekend operation, for example—the 7-day model can be used.

Electronic time clocks offer greater functionality, in particular expanded programming abilities. They can be programmed based on a schedule of any length of time, and automatically adjust operation due to changing sunrise/sunset times, daylight savings, leap year, holidays and special events. This astronomical feature positions electronic time clocks as an effective alternative to photosensors for outdoor lighting control.

Should a blackout or brownout condition occur at the facility, the time clock's internal batteries and backup will ensure it maintains the schedule. In addition, the time clock can stagger the restoration of the loads after a power failure to limit power surges.

Low-voltage Controls

Low-voltage control is a wiring scheme that provides a flexible platform for manual and automatic on/off switching of line-voltage switches using relays (see Figure 2-7).

A simple low-voltage control system consists of a transformer, which steps down line voltage to low voltage; a power switching device that receives low-voltage signals and responds by closing or opening the 20A (or 30A) circuit to switch the lights; and low-voltage wiring used to communicate the control signal. With properly sized wiring, the transformer can feed up to about 15 relays. The on/off switch is wired to the relay using low-voltage wiring, and the relay is wired to the load using line-voltage wiring. When switching from remote locations, pilot lights provide status indication. In this case, because small low-voltage cables replace line-voltage wiring and conduit, this type of remote switching becomes economically viable.

Low-voltage switching is often used to solve complex switching problems. It provides inherent wiring flexibility while also providing the foundation for simple lighting automation. With this foundation, we can begin to add other components, such as control panels and

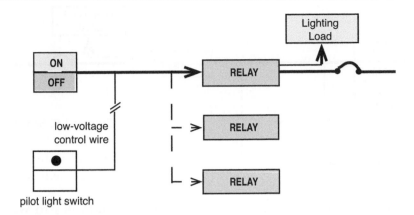

Figure 2-7. Simple schematic for a low-voltage switching system.

photosensors, to add greater flexibility and control of loads.

Digital Switches

The lighting controls industry has been trending towards digital products for years. With low-voltage wiring, digital switches can be used. Digital switches provide the basic functionality of switches but instead of mechanically opening and closing the circuit to switch the load, they send a digital control signal over low-voltage wiring to an occupancy sensor, control panel or power switching device, which switches the load. This allows switches to be networked using a common wiring network, which can result in lower installation costs (see Figure 2-8).

Networking switches allows a simple way to control a single load from multiple switch locations. In addition, digital switches enable the incorporation of greater functionality into the switch, such as multiple buttons that enable on/off switching, timed override and dimming, as well as optional telephone control modules that allow standard telephones to signal the lighting control panel to turn the lighting off.

HID Hi/Lo Switches

Although this book's focus is fluorescent lighting control, it is worth discussing high-intensity discharge (HID) control relative to switching.

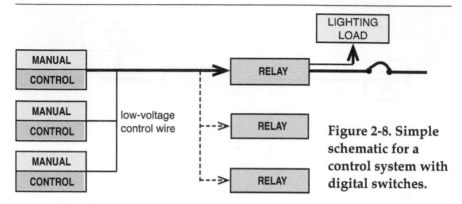

Figure 2-8. Simple schematic for a control system with digital switches.

High-pressure sodium lamps can take 3-5 minutes to warm up, and less than a minute to hot-restrike, reaching full light output in 3-4 minutes. Metal halide lamps take 2-10 minutes to warm up and 12-20 to hot-restrike, while pulse-start metal halide lamps take 1-2 minutes. Given these characteristics, it is often not practical to shut off and restart the lamps based on occupancy if the space must soon be made usable again. Also, most lamp manufacturers rate HID lamp life at a minimum of 10 hours per start. Any reduction in the operating cycle below this minimum will result in shorter lamp life.

Bi-level switching ("hi-lo") ballasts can be an optimal solution for cost-effective automated control of HID lighting. These controls are relay systems that operate HID lamps at either full light output or less (e.g., 50 percent). System components include an on/off contactor, remote bi-level switching ballasts for operation of HID fixtures (also available as a factory-installed option), and input devices.

Occupancy sensors, for example, can be used to switch light levels based on occupancy, with the coverage and control area of each sensor limited to a single aisle. Hi-lo ballasts can respond to control inputs from a variety of accessories, including occupancy sensors, photosensors, timer switches and control panels. For applications that require greater flexibility, dimmable ballasts can be used.

An alternative is to convert HID lighting systems in some hi-bay spaces such as industrial buildings and warehouses to fluorescent high-lumen T8 or T5HO fixtures to save energy and enable greater flexibility in switching based on occupancy, as fluorescent systems start in 0-1 seconds and therefore can be shut off.

Chapter 3

Lighting Control Panels

THE LIGHTING CONTROL PANEL

Lighting control panels—also called switching panels, controllers, sequencers and automation panels—are enclosures that house multiple relays or contactors, a processor that assigns these relays to control zones, and outputs to control the connected loads (see Figures 3-1 and 3-2). Relay-based lighting control panels are essentially input/output devices with low-voltage (24VDC) inputs and line-voltage outputs. Different sizes are available to manage various sizes of loads, from as few as two circuits up to 100+ circuits. The control panel typically is used for control of larger, more complex loads and provides a central platform for lighting automation. Intelligent control panels feature an internal time clock for programmed scheduling. Typical applications include lobbies, corridors, public spaces, retail sales floors, open offices and other applications.

Many office, retail and industrial buildings have been successful in using schedule-based systems as the backbone, supplemented by occupancy sensors and manual switches for smaller offices and special-use areas. The backbone system:

- More easily handles the large amounts of power needed for larger areas

- Switches HID lamps

- Ties into building automation systems where advantageous

Schedules can accommodate the large number of people who share open areas, while allowing people to override the system for special circumstances or emergencies. However, the schedule

Figure 3-1. Control panels. Photo courtesy of Leviton.

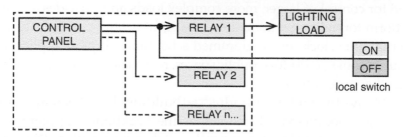

Figure 3-2. Simple schematic for a lighting control panel.

system does not work as well for small areas where the variable work schedule of one person may drive the need for lighting. In those cases, an occupancy sensor or manual switch works well.

Intelligent Control Panels

Intelligent control panels are available with scheduling capability using internal network time clocks. The schedule can be programmed at the panel or at a connected PC (see Figure 3-3). This method enables the panel to become a centralized platform for switching large numbers of loads automatically based on a

Table 3-1. Switching strategies can be used to satisfy a number of project goals in addition to energy management.

Strategy	How It Works	Benefits
Load preservation	Shed non-essential loads during power outages when emergency power is delivered by standby generators	Supports emergency power conditions such as maintaining critical loads
Security response	Control system interfaces with security system to automatically switch lights on in response to intrusion alarm	Supports facility security by supporting response to intrusion
Building system integration	Coordinated switching of non-lighting loads according to a time schedule	Enables management of loads such as fans, heaters, irrigation systems
Research/Horticulture	Tailoring lighting to match seasonal differences in sunrise/sunset	Laboratory research involving animals or plants; plant growth in greenhouses

schedule while still allowing a user to override a particular area for after-hours use. The control panel can also switch in response to an occupancy sensor or a building automation system.

In addition, intelligent control panels can provide monitoring and alarm features. For example, the panels could monitor the status of the branch circuit and various inputs and can alert the facility manager of a tripped breaker, a faulty sensor, or when the burn time of a lamp fixture exceeds a preset value. These alarms can be generated from an embedded email server that allows the operator the ability to click on the email and go directly to the panelboard and view the status from a standard Web browser.

Advantages of intelligent control panels include scheduling for energy savings, code compliance, and individual breaker control, which enables branch circuits to be individually controlled or grouped together and easy rescheduling and changing of control zones via a Web browser. These panels are ideally suited to applications where the granularity of control stops at the branch circuit level, such as retail stores, warehouses, factories, transportation terminals and parking garages.

Figure 3-3. Programmable control panel. Photo courtesy of Square D.

Stand-alone Versus Interconnected Control Panels

While one lighting control panel is sufficient for some buildings, many applications require multiple panels. In these applications, it will either be desirable to operate the panels as stand-alone devices or connect them so that they can share common control signals.

Stand-alone panels: Stand-alone panels are self-contained systems and do not connect to other panels. They typically contain an internal time clock for programmable scheduling of the loads they control, and support their own accessory input devices such as occupancy sensors and photosensors.

Stand-alone panels are advantageous when control functions have to be accessible at each individual panel, and when connecting multiple panels is prohibitive or too costly. Figures 3-4 and 3-5 provide examples.

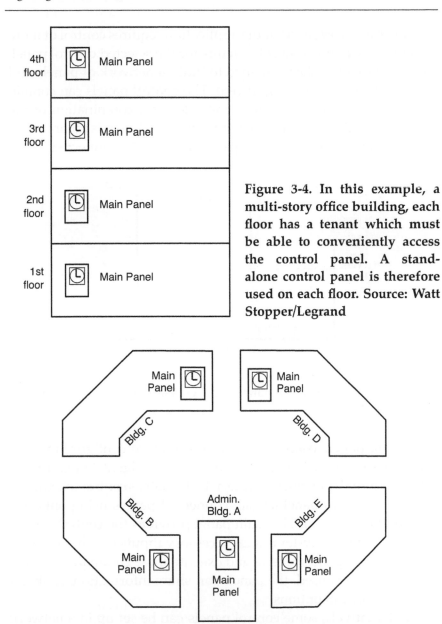

Figure 3-4. In this example, a multi-story office building, each floor has a tenant which must be able to conveniently access the control panel. A stand-alone control panel is therefore used on each floor. Source: Watt Stopper/Legrand

Figure 3-5. In this example, a school campus has several buildings, each served by its own control panel. Because there is no access between the buildings for communications wiring, stand-alone panels are used in each building. Source: Watt Stopper/Legrand

Interconnected panels: If the application requires control of more than the maximum allowable circuits for the selected control panel, then panels can be daisy-chained to build a network of panels and connected controls (see Figure 3-6). The control panels can contain relays (or contactors), dimming modules or a combination of the two for switching, dimming or both functions. There is theoretically no limitation on the number of panels that can be connected.

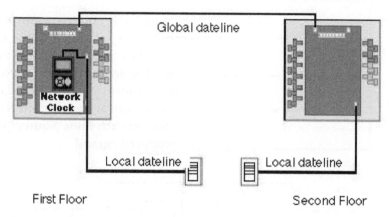

Figure 3-6. Networked panels. Source: Watt Stopper/Legrand.

Using panels connected in a master-slave configuration, the master or main panel can be specified as "intelligent," or equipped with an internal time clock feature, which enables central scheduling for both the main panel and all connected panels using only one clock (see Figure 3-7). The main panel provides the central control signals through its system clock, accessory inputs or other devices. The expansion panels receive and execute these commands. If low-voltage wiring is used, the panels can share information, enabling additional control options.

Alternatively, some control panels can be set up in a network using an open topology, which enables panels to be connected in any configuration, such as a star or T-tap configuration, without needing to plan the wiring as one would with daisy-chained panels.

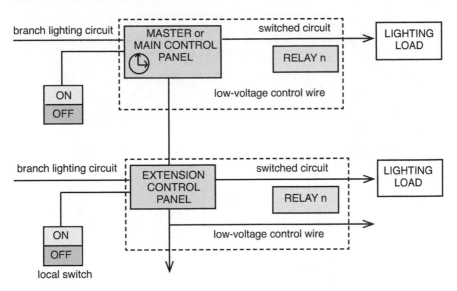

Figure 3-7. Simple schematic for networked lighting control panels.

Figures 3-8 and 3-9 provide examples of applications for interconnected panels.

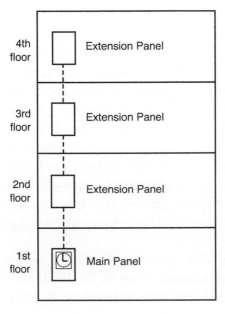

Figure 3-8. In this example, a multi-story office building, the main panel is located on the first floor, where it is convenient for access, operation and routine maintenance by the facility operator. The expansion panels receive and execute commands from the main panel. Source: Watt Stopper/Legrand

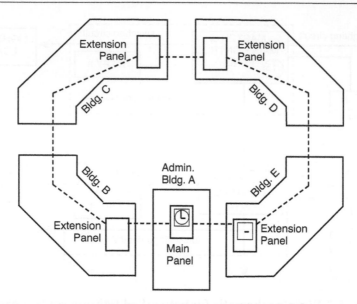

Figure 3-9. In this example, a school campus has several buildings that enables access between the buildings for communications wiring. The main panel is located in the administration building, where it is accessible by the facility operator who can control the lighting across the entire campus from one location. Source: Watt Stopper/Legrand

Centralized Versus Localized Control

In a *centralized control system*, the control panels are located centrally, usually in the electrical room (the lighting control panel may be connected to the breaker panel, or the control panel and breaker panel may be integrated into the same device). The wiring for each switch, and the line-voltage wiring for each load (switch leg), is run back to the panel. In addition, wiring for all accessory inputs such as photosensors must be run back to the panel. This wiring arrangement may not be optimally simple, cost-effective or flexible in applications that require greater control accuracy (smaller zones) or control breadth (multiple accessory inputs). See Figure 3-10.

In a *localized control system* (also called a *distributed control system*), smaller control panels are located closer to the loads they control (and local accessory inputs) and then networked together

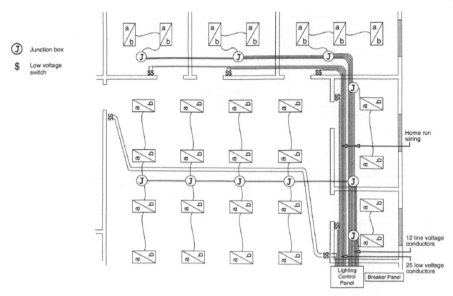

Figure 3-10. Schematic for a centralized control system. Graphic courtesy of Watt Stopper/Legrand.

and to a central time clock, which can be located at a master panel (see Figure 3-11). Basically, localized control entails taking the control panel function out of the central location in the electrical room and dispersing it as smaller panels around the building. The required wiring is reduced to low-voltage twisted-pair communication wiring run between the devices and back to a system time clock located near the electrical distribution panel. This offers advantages for some applications, most notably reduced wiring costs as well as greater simplicity, flexibility and ease of future expansion.

Localized/Distributed control is highly suitable for offices, classrooms, conference spaces and buildings with multiple individual rooms. In particular, localized control is ideal for applications requiring more granular zoning—multiple zones that needed to be switched independently—rather than larger zones with large blocks of lighting switched in unison. It is also ideal for applications with accessory inputs such as occupancy sensors, photosensors, manual switching and signals from a central control point. See Figure 3-12.

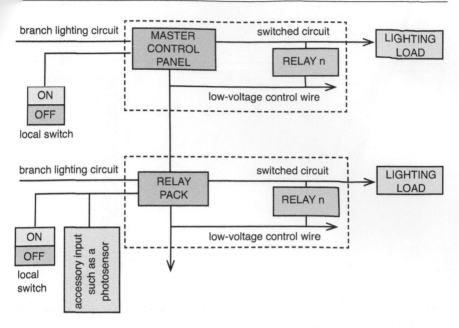

Figure 3-11. Simplified schematic for a localized control system.

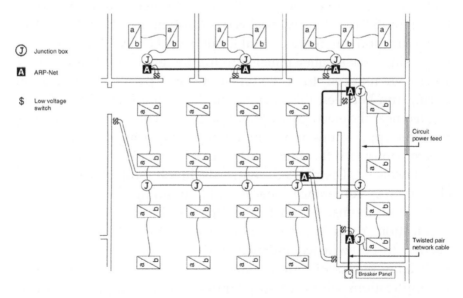

Figure 3-12. Schematic for a localized or distributed control system. Graphic courtesy of Watt Stopper/Legrand.

Localized control is achieved by using small control panels (called automatic relay packs or remote relay packs) that can control two or four separate 20A loads (see Figure 3-13). Typically mounted above the ceiling or in a closet, the relay pack is connected to the load using line-voltage wires and to the switches using low-voltage communication wires. With networked switches, a number of configurations are possible. Signals and commands from PC software, time clocks, other relay packs and other devices are distributed over the communication wires.

Centralized Versus Distributed Intelligence

In a *centralized-intelligence control system*, the processing power is located centrally, such as in a control panel. If the processor fails, the entire control system fails. An example topology for this type of system is expressed in Figure 3-14. The central point represents a central processor, which periodically polls the connected devices for input information that it is programmed to respond to, and then responds by issuing control commands to the connected devices.

Figure 3-13. Intelligent power pack with two relay outputs and two dimming channels delivering both switching and 0-10V dimming control to lighting loads, used for distributed control schemes. Signal inputs offer integrated operation with a range of control devices. Photo courtesy of Watt Stopper/Legrand.

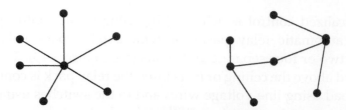

Figure 3-14. Topography of an example centralized-intelligence system (left) and distributed-intelligence system (right).

In a *distributed-intelligence system,* each control device has its own processor, which enables networking of devices using any configuration that the application may require. If the processor fails, the particular device fails but the rest of the system will not be affected, which increases reliability.

Master/Overrides

Remote and local switches can operate together with the panel to allow master control of a floor or department and still allow occupants to override their local lighting. The electronic master switch enables the operator to group and program loads for a degree of central control.

For fluorescent, incandescent and halogen lighting systems (any lighting system that is *not* HID), it is recommended that the lighting controller flick the lights off and on to warn occupants that their area is about to go off.

Another area of concern is that energy codes often require that a local override of a scheduled shutoff event be limited in duration. For example, if a user overrides a scheduled shutoff event and forgets to turn the lights off, then entire control zone(s) could be left on overnight, wasting energy.

To address this, panels can be networked with automatic control switches to provide local override capability to users (see Figures 3-15 and 3-16). These switches are essentially manual on/off switches. However, because the shutoff override may be limited to 2-4 hours or longer depending on the code in effect, these switches can also receive signals from the control panel to turn controlled lighting loads on or off—enabling a timed override.

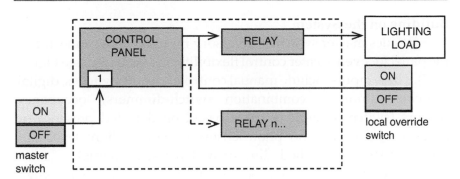

Figure 3-15. Simplified schematic for a control system with master/overrides.

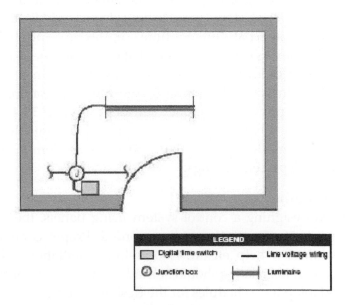

Figure 3-16. Panels can be networked with automatic control switches to provide local override capability to users. These switches are essentially manual on/off switches. However, because the shutoff override is limited to 2-4 hours or longer depending on the code in effect, these switches can also receive signals from the control panel to turn controlled lighting loads on or off—enabling a timed override. Graphic courtesy of Watt Stopper/Legrand.

Building on the System

Besides master switches, accessory inputs can be connected to the panel for even greater control flexibility and accuracy (see Figure 3-17). These inputs include manual controls (on/off switches, digital switches, dimmers, combination switch-dimmers), occupancy sensors (switch lights in a space based on detected presence or absence of people), and photosensors (measures light level and when threshold is reached, dims or switches select lights).

These devices use the processor in the control panel to assign loads to control zones and switch or dim these zones in response to their control signals. The energy savings can add up. The control panel can be scheduled to automatically switch loads at the end of the workday (5-10 percent energy savings have been documented) or as a load-shedding strategy (5-15 percent energy savings).

Through the use of two time clocks, selected light fixtures could be turned on and off at different times or, through split ballasting, different light levels could be obtained at different times each day— a strategy like bi-level switching, but automated on a schedule. Manual dimming provided about 5 percent energy savings in one demonstration project. Occupancy sensors can yield 35-45 percent energy savings. Photosensors operating with dimmable ballasts and dimming modules in the panel have been demonstrated to generate 30-40 percent savings.

Panel Scheduling

When designing a control system using panels, the schedule for each output or circuit must be recorded. Proper completion of the schedule can help ensure trouble-free installation. Changing schedules after installation should be made easy. Many manufacturers provide blank panel schedules for the designer's use.

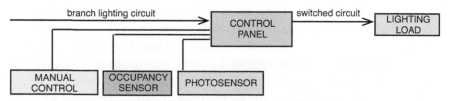

Figure 3-17. Building on the lighting control system.

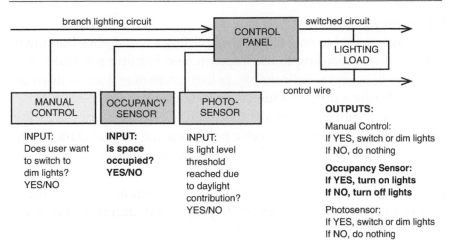

Figure 3-18. Simplified schematic for a lighting control system with multiple inputs to enact multiple energy-saving strategies.

A sample blank panel schedule is shown in Figure 3-19. It requires general panel information, the circuits the panel controls, a description of the controlled loads, workday and weekend (and holiday) schedules for each output, and what controls are designated as able to override the schedule, such as wall switches, telephone or network interface.

Figure 3-19. Generic blank panel schedule.

POWER SWITCHING DEVICE OUTPUTS				TIME SCHEDULE		OVERRIDE CONTROL
Elect Panel	Ckt	Relay	Description	M-F Times	S-S Times	Type, Name, Source
		01				
		02				
		03				
		04				
		05				
		06				
		07				
		08				
		09				
		10				

Emergency Lighting

It is essential to ensure that the building's emergency lighting is not switched according to the implemented lighting schedules, but instead will operate continuously, as in the case of exit signs, or when needed during a power failure, as in the case of emergency units.

Some control panels compartmentalize emergency lighting circuits to ensure they will receive power during a power failure and will operate when needed. Relays can be specified with normally closed contacts to ensure continuation of electrical contact to supply power from a backup power source during a primary power failure. Special emergency lighting remotely operated circuit breakers are also available.

Exterior Lighting

Exterior lights generally fall into two categories. They may be security night lights, which are operated on a dusk-to-dawn schedule. Or they may be general exterior lighting, which are activated at night and shut off during the night when the area is no longer occupied or in use. Exterior lights can be controlled by a time clock, photosensor, occupancy sensor or a combination of these devices. Due to outdoor conditions, you may consider using a more robust device than what you are using inside.

The time clock automatically switches the lights based on the desired schedule. Electronic time clocks allow detailed programming years in advance, and use an astronomical feature to automatically adjust the schedule for changing sunrise-sunset times, daylight savings and leap year. The time clock may also be an internal feature of a control panel. Photosensors measure ambient light level outside and activate or deactivate exterior lights when the light level reaches a target threshold. Occupancy sensors can be used to turn lights on and off based on the detected presence or absence of people, useful for some types of security lighting and to reduce light pollution.

Sample Scheduling Application

Imagine an existing three-story office building with no lighting automation. One of our goals is to enable scheduling to save energy.

Table 3-2. Matching control devices to strategies.

Automatic Shut-off	Individual Space Control	Reduced Light Level Control	Daylight Harvesting	Exterior Lighting Control
Occupancy sensors	Manual switches (on/off switching)	Manual switches (bi-level or multi-level switching)	Manual switches (bi-level or multi-level switching)	Occupancy sensor, photosensor and control panel
Lighting control panel with time clock	Manual switches (bi-level switching)	Occupancy sensors	Photosensor and relay switch (multi-level switching)	Photosensor and control panel
Timer switch	Manual switches (multi-level switching)	Wall-box dimmers	Wall-box dimmers	Control panel with time clock
Building automation system (shut-off, load shedding)	Occupancy sensors	Hi-lo controls for HID	Photosensor, dimmable ballast and dimming controller	Photosensor and control panel with time clock

This will require replacement of the existing lighting control panels with intelligent control panels that feature scheduling functionality. A basic background of the project is provided in Table 3-3.

Table 3-3. Background data for hypothetical intelligent control panel retrofit.

Project	Three-story office building (existing building)
Size	100,000 sq.ft.
Lighting	2x4 fixtures with 4-lamps, 176W each
Power density	2.6W/sq.ft.
Utility cost/kWh	$0.07
Operating hours	5 days/week, 250 days/year
Lighting controls	(2) lighting panels per floor

One of the first steps is to establish baseline conditions so that we can compare options against it. Since we are dealing with scheduling, we need to establish the baseline lighting operating schedule and energy costs before establishing a schedule that is based on occupancy and eliminates waste. This may require a walk-through. In this building, all lights are switched, at panelboard, on @ 6:00 AM by maintenance personnel and off @ 10:00 PM by cleaning crew. See Figure 3-20.

Baseline Energy Cost

Energy = power x time, kW x operating hours, or kWh. We know the load and now know the operating hours, so we can determine our building's lighting energy consumption in kWh. By multiplying this against the utility rate of $0.07/kWh, we can determine our buildings lighting energy cost per day and per year.

Estimated workday average hours of lights on (normal work hours, after hours clean-up and maintenance) = 16 hours.

Cost per day = 16 hours x ($0.07/kWh x 2.6W/sq.ft. x 100,000 sq.ft.)/1000

Cost per day = $291.20.

Annual cost = $291.20 x 250 days/year.

Annual cost = $72,800.

Now that we understand the baseline conditions, we can choose a new control strategy to compare against it. In our building, the lighting panels will be replaced with six 14-circuit intelligent lighting control panels that enable scheduling of select loads. All

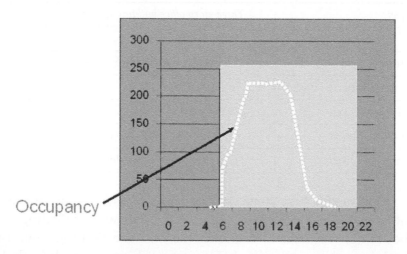

Figure 3-20. Baseline operating schedule. Graphic courtesy of Square D.

lights will be switched according to the automatic control system (see Figure 3-21):

- 100% on @ 6:00 AM
- 75% off @ 6:00 PM
- 100% off @ 8:00 PM (30-minute cleaning crew override provided per panel, 15% load for 3 hours after 8:00 p.m.)

See Figure 3-21.

Projected Energy Cost
100% load for 12 hours (6 am - 6 p.m.):
Cost per day = (12 hours)($0.07/kWh x 2.6W/sq.ft. x 100,000 sq.ft.)(1.00)/1000.
Cost per day = $218.40.
25% load for 2 hours (6 p.m. - 8 p.m.):
Cost per day = (2 hours)($0.07/kWh x 2.6W/sq.ft. x 100,000 sq.ft.)(0.25)/1000.

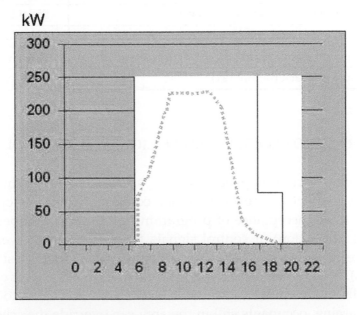

Figure 3-21. New operating profile with intelligent control panels. Graphic courtesy of Square D.

Cost per day = $9.10.

Cleaning crew override:

Cost per day = (3 hours)($0.07/kWh x 2.6W/sq.ft. x 100,000 sq.ft.)(0.15)/1000.

Cost per day = $8.20.

Annual operating cost (250 business days/year):

Annual cost = ($218.40 + $9.10 + $8.20)(250).

Annual cost = $58,925.

Projected Operating Cost Savings

Our replacement of the existing control panels with intelligent control panels is estimated to pay for itself in less than two years, after which it can deliver about $14,000 cash flow per year resulting from lower operating costs (see Table 3-4). This cash flow translates to profits, and greater profitability and competitiveness for the owner.

Table 3-4. Projected operating cost savings and payback.

Energy cost (existing panels)	$72,800
Energy cost (new control system)	$58,925
Estimated annual energy savings	$13,875
Estimated installed cost for new control system	$27,000
Simple payback (before tax)	1.95 years

CENTRALIZED PROGRAMMABLE LIGHTING CONTROL

Lighting control panels can be specified with an internal modem and telephone or Ethernet connections. The telephone input enables acceptance of programming from facility operators (global instruction) or control signals from users (local override) using telephones, cell phones, beepers or personal digital assistants (PDAs). For example, if the control system is about to shut off the lights in an area due to a schedule, and causes the lights to blink as a warning, occupants still in the area can override the schedule and extend the lighting period by simply picking up the phone and punching in several numbers.

The Ethernet input enables acceptance of operation and programming of the panel/lighting network from authorized PCs either on the building local area network (LAN) or located remotely and communicating via the Internet. In this case, the building LAN is used to control lighting instead of low-voltage wiring. This provides single-point control system programming and modification, combined with threshold-based local controls, of any number of loads in a space, building or campus. Other ports can be specified in the controller for acceptance of other control signals such as DMX512 signals from theatrical control systems.

Figure 3-22 shows a centralized lighting control system using centralized intelligence and PC-based scheduling and programming.

Networked lighting control systems communicate with each other and with a central terminal, usually a PC. The facility operator can program schedules into the PC's control software and modify them later based on changing conditions. Networked lighting control systems provide three main incremental functions: central programming/monitor/control, global commands, and management data.

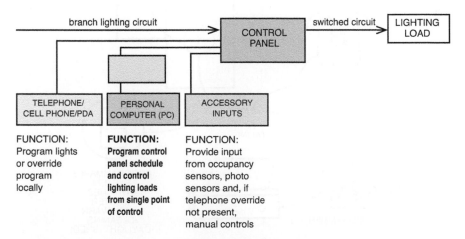

Figure 3-22. Centralized lighting control system using centralized intelligence and PC-based scheduling and programming.

While there many configurations for networked control systems, an example of a distributed panel system is shown in Figure 3-23. It provides cost-effective automated lighting control for applications ranging from a small office building to a mall to an industrial complex. Each of the distributed control panels has stand-alone automation capability. The network links these controllers to a central operator terminal (PC). In order for the network to differentiate between devices, each must have a unique address or identification.

Lighting control networks allow the central collection of operating data and status information for measurement of energy consumption, load profiling, and building management functions. For example, at the end of the month, the operator can simply ask the system for a report on the total lighting energy consumption for the last period. If that consumption is excessive, the operator

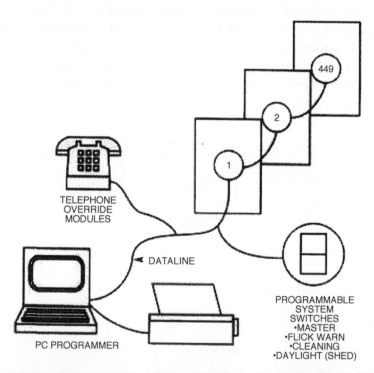

Figure 3-23. Distributed panel networked control system. Graphic courtesy of National Electrical Manufacturers Association (NEMA).

then might ask for a report of every load that exceeded its expected runtime during the month. Having identified the "offenders," a profile of the actual runtime for each can be used to identify why and how the excess occurred. Such management data is critical to ensuring that automated lighting systems continue to save energy. In addition, this same information can become the basis for a fair allocation of lighting costs by tenant or department.

In addition, an intelligent system can also provide alarm functions, such as automatic notification of tripped breakers, alarm on faulted components, and alarm when branch circuit run-time exceeds the preset value.

Figure 3-24 presents a centralized programmable lighting control system with centralized intelligence.

Building Automation Systems

A microprocessor-based centralized programmable lighting control system is basically a microprocessor-based centralized load controller. Although it is designed principally for lighting, the system can handle HVAC, service water heater, and motor loads.

Another possibility is to use a building automation system (BAS) or energy management system (EMS). Building automation controls are capable of controlling lighting systems as well as HVAC, security and fire safety systems. Depending on the options specified, they can perform many other functions as well, such as maintenance scheduling, monitoring, logging and inventory control. In addition, they can be integrated with energy information systems (EIS), which can provide a continuous flow of information, such as real-time changes in energy prices, which can be used by the EMS to automatically adjust the performance of the control systems.

The approaches used for lighting control by a building automation system are essentially similar to those associated with centralized programmable lighting control systems. Lighting systems can be integrated easily and virtually all the different functions described earlier can be controlled from one central location, relying on the appropriate sensors, actuators and monitors, connected by multiplexed transmission media.

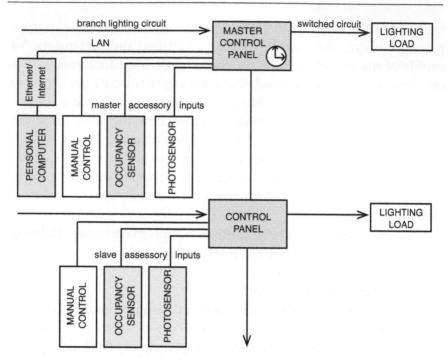

Figure 3-24. Simple schematic of a centralized programmable lighting control system with centralized intelligence.

For example, when a person enters an area, an occupancy sensor turns on the lighting, which causes the HVAC system to respond to provide conditioned air to that area. An icon on the building automation system's workstation screen changes color to show the change in status. The building automation system's operator controls and monitors the building's lighting, HVAC, fire protection and access controls from a single workstation.

For the lighting control system to be able to communicate with the building automation system, a common language is needed. The most popular are BACnet and LonWorks. Devices designed according to these protocols can interact seamlessly with similarly designed devices. Or a gateway can be used, which is a device that acts as an interpreter between controls built to operate on different protocols.

Part II

Dimming
Fluorescent Systems

Chapter 4

Dimmable Ballasts

Fluorescent dimming has gained popularity in recent years as a control strategy to save energy and increase the range of light levels provided by a lighting system.

Fluorescent dimming can enable energy cost saving strategies such as daylight harvesting and demand limiting, provide greater flexibility to adapt the lighting system to changing space needs, meet visual needs by controlling the quantity of light in the space, and provide personal control to occupants, which can increase worker satisfaction.

Potential benefits of dimming controls include:

- Improved aesthetics and image
- Greater space flexibility
- Mood setting
- Energy and other utility cost savings
- Increased worker satisfaction and productivity
- Sustainability—less pollution by power plants
- Enhanced space marketability
- More effective facilities management

Additionally, with more stringent lighting design guidelines, lighting designers have the freedom to incorporate fluorescent fixtures into their design in order to achieve the desired watts per square foot without compromising ambience in the space.

Note that dimming systems add value but also add cost to a project. Both are primary considerations in design planning. Additional costs may include controls, labor and materials for additional wiring if low-voltage dimming methods are used, and a premium for dimmable ballasts. The premium contributed to the average project cost can be about $0.75/sq.ft. for the dimmable ballast and about $1.00/sq.ft. for a photosensor and additional control wiring (if low-voltage). Costs, however, have been declining.

Fluorescent lamps were not originally designed to be dimmed by ballasts. The first dimming ballasts, which were electromagnetic in design, offered a limited dimming range with a number of tradeoffs such as flicker and other problems.

Today's dimmable electronic ballasts generally provide efficient, reliable performance without the tradeoffs associated with dimmable magnetic ballasts.

In this chapter, you will learn about how dimming ballasts work, how they fit into a system, what methods are used, ballast types, lamp types and dimming range. This chapter focuses on electronic dimmable ballasts, which have become most popularly used.

Figure 4-1. These light fixtures are fitted with dimmable ballasts that respond to user commands to dim to their preferences. Photo courtesy of Ledalite Architectural Products.

Table 4-1. Sample applications and realized benefits of fluorescent dimming.

Space Type	Benefit
Discount Retail Store	In an open retail space with daylighting, dimming can reduce electric lighting use but allow the lights to be on, making the store seem "open for business."
Conference Room	Dimming lighting can facilitate a variety of visual presentations.
Health Care Facility	Daylight-driven dimming can provide a smooth and unnoticeable transition to electric lighting as daylight levels decrease, while maintaining the desired light level. Also provides the added benefit of night-time patient care with low illumination levels.
Restaurant	Preset scene dimming controls can make changing the ambiance as the day goes on consistent and as convenient as pressing a button.
Office Area	Even in an open office area, occupants can be given the option of dimming the light fixture over their workstation to suit their personal preferences.

Table 4-2. Energy management-focused dimming strategies and potential energy savings.

Strategy	Description	Potential Energy Savings
Tuning	This entails providing manual dimming control of localized lighting so that each user can "tune" the lighting to their personal needs and preferences. This strategy can increase worker satisfaction and productivity, while generating energy savings.	5-20%
Scheduling	This entails automatic dimming and/or switching based on a programmed schedule. This strategy can generate significant cost savings.	10-30%
Load shedding	The entails acquiring the ability to reduce lighting power in response to a signal from the utility. This allows the utility to respond to heavy demand without adding generating capacity. The owner receives a special rate in return. Load shedding may also be used to reduce the loading on the air conditioning system during a heat wave. This can allow a reduction in the size of the air conditioning system, offsetting the cost of the dimming system.	Demand and energy savings

(Continued)

Table 4-2. Energy management-focused dimming strategies and potential energy savings (*Continued*).

Strategy	Description	Potential Energy Savings
Daylighting	Good daylighting uses natural light to provide a well-lit space without introducing glare or drastically diminishing the quality of the visual environment. Energy savings are realized when the electric lighting is dimmed or switched. This strategy can generate significant cost savings by taking advantage of "free" light from the sun.	30-40%
Adaptive compensation	This entails reducing light levels at night in spaces with non-critical tasks based on research that people prefer and need less light at night than during the daytime. The reduction should be done gradually. When an outside photosensor detects a drop in external light level, the lights are dimmed accordingly. This strategy can improve lighting quality while generating significant cost savings.	Up to 50%
Lumen maintenance	The light output of some lamps decreases over time. This is taken into account by lighting designers in their plans, generally meaning the lighting system will be over-designed to produce more light than is needed initially. Lumen maintenance dimming uses a photosensor to maintain a consistent light level throughout the lamp's age and performance, thereby saving energy—up to 5-10 percent. Due to large-scale improvement in lumen maintenance by lamp manufacturers, however, this strategy's viability has been significantly decreased.	5-10%

Theory of Operation

Fluorescent ballasts provide proper starting voltage to initiate an arc of electric current between the lamp cathodes, and then regulate the current flowing through the lamp (see Figure 4-2).

The ballast can be designed so that it 1) receives a signal from a dimmer as an input and subsequently 2) changes the current flowing through the lamp as an output.

Changing the current will result in a reduction in lumen output and drawn power. The control signal's characteristics affect the duration and extent of the change in current and subsequent lamp output and power.

During dimming, the ballast must maintain a sufficiently high voltage across the lamp to sustain the arc of current flowing through the lamp, and keep the cathodes properly heated.

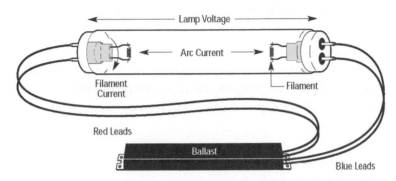

Figure 4-2. Fluorescent lamp dimming is achieved by changing the current, which results in a reduction in lumen output and drawn power. Graphic courtesy of Lutron Electronics.

Dimming Methods

The dimmable ballast is essentially an input/output device. The ballast receives input (ballast and controller communication through control signals) and responds with output (alteration of lamp current and subsequent wattage and light output). How the ballast does this is determined by its dimming method.

A number of dimming methods are available, including analog (step dimming, 0-10VDC, two-wire phase-control, three-wire phase-control and wireless infrared) and digital. Essentially, the control circuit can be in the form of low-voltage leads from the ballast (0-10VDC, digital), incorporated into the line circuit (phase-control), or wireless (wireless infrared).

The next chapter covers dimming methods in detail. Selection of dimming method/ballast is typically based on the use of the space, dimming range, wiring, lamp type, physical size and/or budget.

A Fluorescent Dimming System

A typical fluorescent dimming system will include (see Figure 4-3):

- The power source
- Dimming controller
- Fluorescent lamp(s) (any type of fluorescent lamp except 25W, 28W and 30W T8 lamps and some larger compact fluorescents)
- Dimmable ballast
- Control (communication) wiring
- A light fixture featuring rapid-start lamp sockets
- Optionally other input devices such as occupancy sensors and photosensors

All of these components must be compatible for the system to operate as an integral whole.

A good example of a typical system is 0-10VDC. This is the most popular dimming method. In addition to two leads used to connect the ballast to the power supply, 0-10VDC ballasts feature two additional low-voltage leads that control lamp output.

Figure 4-3. Components of a basic fluorescent dimming system. Graphic courtesy of Advance.

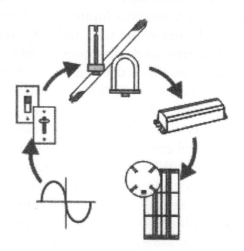

These leads form a two-wire low-voltage bus to which controllers can be connected. For example, if a photosensor is connected to the bus, it can vary the voltage on the bus and therefore ballast output. Typically, a number of ballasts are connected in parallel on the bus so they can be controlled as a group from a single photosensor or controller.

Other ballasts operate on the same principle; however, they operate with either a two-wire or three-wire topology (line-voltage wiring method) and not 0-VDC (low voltage wiring method).

Dimming Range

Dimming is often described as a percentage, such as 20 percent dimming. This percentage can refer to lamp current (A), lamp voltage (V) or lamp power (W), although it usually refers to relative light output and is then related to light level.

For example, a 1000-lumen lamp at 20 percent dimming would produce 20 percent of its light output, or 200 lumens.

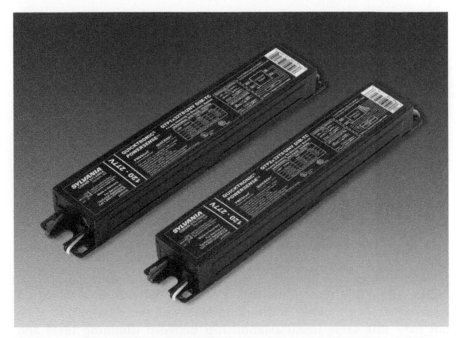

Figure 4-4. Dimmable ballasts. Photo courtesy of OSRAM SYLVANIA.

While most dimmable electronic ballasts operate with a linear relationship between dimming level and lamp output, they do not have this same linear relationship between dimming level and power—e.g., a 20 percent dim level will result in 20 percent of the original lumen output but not 20 percent of the original wattage (see Figure 4-5).

Table 4-3. Dimming range and typical applications.

Application Goal	Available Dimming Range
Energy management	100% to 5-10%
Visual needs	100% to <1%

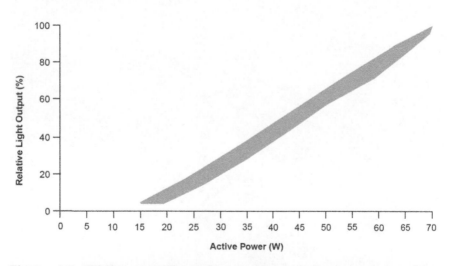

Figure 4-5. While most dimmable electronic ballasts operate with a linear relationship between dimming level and lamp output, they do not have this same linear relationship between dimming level and power. Graphic courtesy of the Lighting Research Center.

Linear lamps: Dimmable electronic ballasts feature feedback circuits that maintain voltage to the electrodes during a reduction in lamp current. This allows dimming without causing the lamp to extinguish. It also allows continuous dimming over a wide range without reducing service life.

Dimmable electronic ballasts for linear lamps can be categorized as either:

- *Energy management application ballasts* with a range from 100 percent to 5-10 percent of light output. An example of an energy management application is an open office plan where electric light output is reduced by say 20 percent each day during the utility peak demand period. In this case, dimming allows us to reduce utility costs with minimal disruption to work operations. See Table 4-4.

- *Premium architectural dimming application ballasts* with an available range from 100 percent to 1 percent of light output. An example of an architectural dimming application is a conference room, where occasionally we will want to achieve very low light levels for audio/video tasks. In this case, the dimming itself provides the value, with energy savings being a by-product. See Table 4-4.

Dimming range is just one area where the capabilities of dimmable electronic ballasts have been extended in recent years. Just a few years ago, the limit for the range of energy management ballasts was 20 percent, but now it is 5-10 percent. While the effective floor for energy savings purposes is 20 percent, as no significant savings are achieved below that level, the newer generation of dimmable ballasts provides greater flexibility with the expanded range.

Note, however, that a number of factors may affect actual dimming range in a field installation. Dimming range can vary based on voltage to the ballast, ambient temperature, initial lamp burn in, and lamp life used/remaining.

CFLs: Dimmable electronic CFLs are available with a dimming range from 100 percent to <3 percent.

Magnetic ballasts: Dimmable magnetic ballasts for operation of linear lamps operate differently than dimmable electronic ballasts. Whereas electronic ballasts maintain lamp voltage during reductions in lamp current, magnetic ballasts reduce input power to the ballast to reduce the lamp current, which also reduces the voltage to the lamp electrodes. As a result, the dimming range is limited to 100 percent to 20 percent.

Table 4-4. Dimming range for various dimmable ballast types.

Dimming Method	Applications	Available Dimming Range
Dimmable electronic ballasts for linear lamps	Energy management	100% to 5-10%
Dimmable electronic ballasts for linear lamps	Architectural dimming	100% to 1%
Dimmable electronic ballasts for CFLs	Energy management, mood-setting	100% to <3%
Dimmable magnetic ballasts for linear lamps	Energy management	100% to 20%
Fluorescent step dimming	Energy management	Usually two or three levels: 100% and 50%; 100%, 66%, 33%, 0%

Lamp Types

Dimmable electronic ballasts are available in:

- One-, two-, three- and four-lamp models operating linear rapid-start T8 lamps

- One- and two-lamp models operating linear and twin-tube T5 and T5HO lamps

- One- and two-lamp models operating 4-pin compact fluorescent lamps (CFLs)

The continuous-dimming range may differ based on the type of lamp. For example, the dimming range for a 0-10VDC ballast operating F32T8 lamps may be 100 percent to 3 percent, while the dimming range for a 0-10VDC ballast operating F54T5HO lamps may be 100 percent to 1 percent.

As of April 2007, dimming was not available for "energy saving" (25W, 28W and 30W) T8 lamps and higher-wattage (60W, 85W and 120W) CFLs.

Lamp Sockets

Linear lamps operated by dimmable electronic ballasts feature bi-pin bases typical of rapid-start lamps. Dimmable systems therefore require the use of rapid-start lamp sockets that accept two distinctly separate wires for each side of the lamp socket.

In contrast, instant-start sockets are shunted on the one side to prevent acceptance of the second ballast wire or contain a shunt wire, and are therefore not compatible with dimming ballasts. See Figure 4-6.

Note that lamp life can be reduced if the lamp filaments do not make proper contact in the socket. It is often recommended to use knife-edge sockets rather than flat-edge sockets for operation of linear T8 and T12 lamps.

Figure 4-6. Linear lamps operated by dimmable ballasts must have rapid-start sockets that accept two distinctly separate wires for each side of the lamp socket. Instant-start sockets, in contrast, are not compatible with dimmable ballasts. Graphic courtesy of Leviton.

Electronic Versus Magnetic Ballasts

Dimmable magnetic ballasts are available, but are prone to lamp flicker, premature lamp failure when dimming at the low end for long periods of time, color shift at the low end, lower efficiency and a dimming range of 100 percent to 20 percent. For years, the choice for dimming in fluorescent applications was either to dim magnetic ballasts and accept these tradeoffs or to add a layer of incandescent lighting into the space for the purpose of dimming. Due to these factors, dimmable magnetic fluorescent ballasts have become quite rare.

Today, most dimmable ballasts are electronic ballasts. Dimmable electronic ballasts operating linear lamps in a properly specified, installed and operated system will not experience the previously described problems commonly encountered with dimming magnetic ballasts. Electronic ballasts are more efficient, and offer a maximum dimming range, depending on the ballast type, of 100 percent to 1 percent. They also operate more quietly, are smaller and lighter in weight, and can be designed to dim a wide selection of wattages and lamp types among linear lamps and compact fluorescent lamps.

Early dimmable electronic ballast models experienced failures and problems just as fixed-output electronic ballasts did, but like fixed-output ballasts, since the days of early adoption the technology has matured and achieved a high degree of reliability and an ongoing evolution of capabilities. As stated earlier, this chapter focuses on electronic ballasts, as today these are the most commonly installed type of dimmable ballast.

Programmed-start Ballasts

To optimize lamp life, the ballast must be able to provide adequate cathode heating at startup and during operation throughout the full dimming range. For this reason, most dimmable ballasts use a rapid-start circuit, since instant-start circuits provide no cathode heating. Instant-start systems are more efficient—i.e., require fewer watts at full light output—than rapid-start systems, although new dimmable ballasts are available that demonstrate comparable efficiencies of standard instant-start ballasts.

Older dimmable electronic ballasts are rapid-start, but most of today's dimmable ballasts are programmed-start rapid-start ballasts that have an on-board microprocessor that reads electrical characteristics of the lamp and adjusts its performance accordingly.

Programmed-start ballasts are rapid-start ballasts that preheat the electrodes more accurately to minimize damage to the electrodes during the startup process (according to a program) and therefore can optimize lamp life. As a result, programmed-start ballasts can provide up to 100,000 starts with no negative effects to the manufacturer's rated lamp life, highly suitable for frequently switched applications such as spaces controlled by occupancy sensors.

Additional
Compatibility Issues

It is not recommended to mix different fluorescent loads on a dimmer (like incandescent with magnetic or electronic ballasts). There is an increased possibility of bad performance, along with more complicated troubleshooting.

Linear lamps and CFLs may have different dimming ranges. It is therefore recommended to group each lamp type on a separate dimming control zone or channel.

Non-dimmable (fixed light output) linear and CFL ballasts should not be installed on circuits controlled by a dimmer. Select a switch or non-dim module for these loads when designing a dimming system.

Specify ballasts that are rated by the manufacturer as compatible with control devices that use the same dimming method.

Note that some manufacturers of 0-10VDC ballasts provide command regions in the 0-10VDC range; a signal less than 0.3V might signal the ballast to shut down, for example. Be sure that the specified controllers are compatible with any such added feature for the chosen ballast.

Specification Tips

Avoid mixing loads on the same dimmer: It is recommended to avoid mixing different loads on the same dimmer, such as incandescent lamps and fluorescent ballasts. With dimming systems, different loads can be mixed on the same control station, each with their own zone, with the use of a different dimmer at the panel designated for each load.

Operate ballast within parameters: Note that some devices can be dimmed only to a specific level. While some ballasts may provide lamp ignition at any level, some may also require a minimum turn-ON level. Ensure that the ballast is not dimmed or turned on below these levels. Some manufacturers provide a low-end trim feature to avoid this.

Account for voltage leak: Many dimming systems leak voltage to the ballast when "off," which can cause the ballast to fail prematurely and also cause lamp flicker. It is recommended that the dimmer be positively disconnected from power when it is off. This may be accomplished with a relay in line. Consult the controls manufacturer.

Account for de-rating: The current-carrying capacity of dimmers may be expressed in amps, watts, ballasts, etc. and may be also based on the voltage of the AC power supply. It is recommended that circuits feeding ballasts should not exceed 80 percent capacity and should be de-rated accordingly. Because of large in-rush currents, magnetic ballasted loads should be de-rated up to 50 percent. De-rating may also be required for ganging dimmers.

Account for in-rush current: When most lighting loads are switched on, they can draw a much higher level steady-state current for the first couple of line power cycles—a fraction of a second. In some electronic ballasts, particularly ones using front-end active filters to reduce harmonic distortion, this in-rush current may cause the contacts in lighting relays to fuse and damage connected devices such as dimmers. It is recommended to size any lighting relays to account for this. In addition, some dimmable ballasts feature in-rush current-limiting circuitry. NEMA Guideline 410-2004 sets voluntary guidelines for manufacturers of lighting controls and switching devices with electronic ballasts; look for ballasts that meet these guidelines.

Ensure compatibility and interoperability: Be sure that the selected ballast is dimmable, that the ballast is rated to operate the selected lamp, and that the ballast and controls are compatible. To ensure consistent dimming performance, it is recommended that all lamps controlled by a dimmer should be of the same brand and wattage. (Note, however, that many current dimmable ballast designs allow for similar operation of 17W, 25W and 32W lamps on the same circuit, typically in cove lighting applications.) Similarly, all ballasts controlled by a dimmer should of the brand and model.

Account for ballast power draw in off state: Some dimmable ballasts are designed to switch the lights completely off after reaching the low end of their dimming range, but will continue to draw power while off. This should be accounted for in energy calculations. Consult the manufacturer.

Stay within maximum ballast lead lengths: The ballast manufacturer will provide a maximum length for the leads connecting the ballast and the lamp. Exceeding these maximum lengths can result in poor dimming performance and early lamp failure.

Stay within maximum control wire lengths: If the dimming system is based on the 0-10VDC dimming method, the control manufacturer will provide a maximum length for the control wires running between the dimmer and the ballast. To prevent dimming performance being affected by electrical noise or interference, consider conduit or shielded cable if the wiring run length must be extremely long and/or electrical noise or interference is present. This may also be a code requirement in some regions. Note that digital ballasts also use low-voltage control wiring but is not subject to noise or interference.

Dimming and Light Level Perception

An important design factor related to dimming is perceived brightness versus actual light level. As lamps are dimmed, light output decreases but the human eye may perceive a higher light output and light level than is actually present. This is because the human eye overcompensates for diminished light level by allowing more light to enter into its pupil. For example, dimming to 25 percent appears to be about 50 percent of the original light level. The effect is predictable according to the square law, which defines the theoretical relationship between light level and perceived brightness (IES *Lighting Handbook*):

Perceived Light (%) = 100 x square root (Measured Light (%)/100)

Understanding the relationship between dimming and perception of light level is very useful in planning dimming systems. Assume an installation in which the measured light level is 60 fc at full light output. The goal is to dim the lighting system as low as possible while maintaining a perceived light level of at least 10 percent.

- 1 percent measured light (0.6fcd) is perceived as 10 percent (desired result)

- 5 percent measured light (3fcd) is perceived as 22 percent (2x brighter than desired)

- 10 percent measured light (6fcd) is perceived as 32 percent (3x brighter than desired)

While multi-level fluorescent switching strategies are generally more economical (from an initial cost point of view) than continuous dimming, a disadvantage is that occupants often regard sudden changes in light levels as disruptive. For this criterion, continuous dimming is superior. But do occupants notice the change during dimming, and if so, at what point?

The Lighting Research Center studied the relative threshold for detection of gradual reduction in light levels via continuous dimming. Four sessions were conducted. Sessions A and B were conducted in a room with more than twice the light level of Sessions C and D. The A, B curve shows:

- More than 90 percent of the population would not notice a 10 percent reduction in light output.

- About 75 percent would not notice a 15 percent reduction in light output.

- About 55 percent would not notice a 20 percent reduction in light output.

The Lighting Research Center concluded that these results can be considered a maximum since the subjects in the experiment were aware that the light level was about to change, which does not match real world conditions (see Figure 4-7).

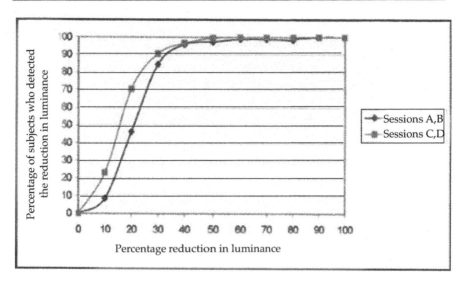

Figure 4-7. Results of Lighting Research Center study concerning relative threshold for detection of dimming. Graphic courtesy of Lighting Research Center.

Table 4-5. Is dimming needed at all for energy management applications, or are on/off strategies such as multi-level switching sufficient?

Strategy	Dimming	Multilevel Switching
Approach	Smooth continuous reduction in light levels over wide range	Stepped reduction in light levels by switching select lamps and ballasts
Advantages	Greater flexibility in selecting light level; smooth transitions between light levels; greater user acceptance	Less expensive and reduced often higher energy savings commissioning—dimming typically adds $0.50 to $1.00 per square foot to the installed cost
Disadvantages	More expensive and increased commissioning required	Abrupt changes in light levels can be disruptive in spaces dominated by stationary tasks; less flexibility; often less energy savings due to less flexibility
Applications	Visual needs applications such as conference rooms, training rooms, auditoriums; energy management applications with users performing stationary tasks	Energy management applications such as offices, classrooms and intermittently occupied spaces such as hallways; the more steps, the more accepted and the higher the energy savings

Chapter 5

Fluorescent Dimming Methods

Fluorescent dimming systems can satisfy the visual needs of the space, change light levels without disruption to operations, and reduce utility costs through daylight harvesting, demand reduction, scheduled dimming and other strategies.

Dimmable ballasts are an essential building block of fluorescent dimming systems and, like other electrical devices, can be viewed in terms of useful inputs and outputs. Inputs include control signals, and outputs include reductions in lumen output and power. To perform this function, dimmable ballasts must be configured to understand and act upon the control signal coming from a control device, and then act upon the current flowing through the lamp to achieve the dimming effect.

How these functions are performed is called the dimming method.

This method becomes the basis for design, construction and operation of dimmable ballasts and associated control devices. Ballasts and controllers must be built on the same method for them to be interoperable and able to communicate. Just as dimming and non-dimming equipment should not mix on the same zone or circuit, equipment designed to operate using different methods are not compatible and should not be operated in the same system. When incompatible dimmers and controls are specified or installed, the result can be incorrect wiring, poor dimming performance, or damaged equipment.

There are a number of methods available for fluorescent dimming. Some are open to the industry as a non-proprietary technology, while others are to an extent proprietary to a specific manufacturer. In this chapter, you will learn about the types of available dimming methods and their performance characteristics. Specifically, you will learn about analog methods including 0-10VDC (see Figure 5-1), two-wire phase-control, three-wire phase-control and wireless infrared. You will also learn about the other primary dimming method—digital dimming, and the DALI protocol.

Analog versus Digital

The primary methods used to dim fluorescent lighting include analog and digital. Analog dimming systems are established and common, while digital dimming systems are relatively young in the industry. Both provide the essential function of controlling the lamp output based on input from a control device. Both enable the construction of networks of controls and ballasts wired to local and central points where control signals can originate, either manually or automated.

Analog typically presents a lower component cost than digital and is compatible with a wide range of common control devices. The dimming ballasts can be on a low-voltage or line-voltage control circuit. Analog ballasts and controls are compatible as long as they are configured to the same method. There are several analog methods, including 0-10VDC, two-wire phase-control, three-wire phase-control, wireless infrared and wireless, with 0-10VDC, the

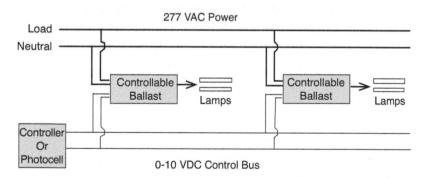

Figure 5-1. The most popular type of dimmable system for energy management applications utilizes the 0-10VDC analog method. Source: California Energy Commission.

oldest of these methods, being most popularly used.

Interoperability

While there are a number of industry-adopted analog methods, there is currently no industry standard for operation of analog dimming ballasts. Manufacturers of 0-10VDC ballasts, for example, will publish lists of manufacturers whose 0-10VDC controllers are compatible with their ballasts. Ballasts and controllers built on the same method are interoperable in the same dimming system.

However, dimming performance may not be consistent among ballasts made by different manufacturers and across different ballast types. A 5V signal might result in a 50 percent dimming level on one manufacturer's ballast but 30 percent on another manufacturer's ballast. For this reason, it may be desirable to avoid mixing ballast types from different manufacturers in the same dimming system so as to ensure consistent performance.

ANALOG 0-10VDC DIMMING METHOD

The 0-10VDC dimmable electronic ballast includes components that perform these functions: electromagnetic interference filtering, rectification, power factor correction and ballast output to power the lamp.

When the voltage level is near or above 10VDC, the ballast responds with full light output. As the voltage decreases, the ballast reduces light output. Note that some manufacturers provide command regions in the 0-10VDC range; a signal less than 0.3V might signal the ballast to shut down, for example. Be sure that the specified controllers are compatible with any such added feature for the chosen ballast.

The ballast can also be connected to a switch or relay to enact bi-level control, providing full light output when the switch opens and reducing it to a specified minimum when the switch closes.

Dimming is accomplished by controlling the amount of current flowing through the lamp by reducing the lamp power. As lamp power decreases, lamp voltage increases proportionally to maintain heating of the lamp cathodes and prevent the lamp from being extinguished.

A dimming range of 100 percent to 3 percent ballasts is available for T8 lamps and CFLs and 100 percent to 1 percent for T5HO lamps.

Ideally suited for energy management applications, both new construction and retrofit, 0-10VDC systems are well suited for auditoriums and training areas, conference rooms and boardrooms, department and specialty stores, education, healthcare, hotels, houses of worship, private and executive offices, restaurants and other spaces.

Wiring

0-10VDC is a low-voltage wiring method with four wires, with two line-voltage leads (hot and neutral) to power the ballast and two low-voltage control leads to change the light level. Typically, 0-10VDC ballasts have violet and gray control wires. The gray wire is internally connected to provide a ground reference. Many new-generation dimmable electronic ballast lines include dual- or multi-line voltage models and provide overvoltage protection of low-voltage leads in the event that line voltage is accidentally applied.

Depending on wire insulation and control switch ratings, the control wires may either be routed in the same raceway as the power wires (Class 1) or in a separate raceway from the power wires (Class 2). In general, the system may be installed as Class 1 if the control wires carry the same voltage rating as the power wires and the control device is rated for Class 1. See Figure 5-2.

Conduit may be required for the low-voltage wiring, however. In installations with long wire runs, it is recommended that the 0-10VDC wire be placed in conduit in order to separate the low-voltage wiring from line-voltage wiring. When low-voltage wiring is run next to line-voltage wiring, over long distances and without conduit, voltage may be induced onto the low-voltage wiring, resulting in loss of the control signal.

Class 1 installation

Class 2 installation

Figure 5-2. 0-10VDC control wires may be routed either in the same raceway as the power wires (Class 1) or in a separate raceway from the power wires (Class 2). Graphic courtesy of Advance.

Assessment

0-10VDC ballasts are the most popular dimmable electronic ballast type and are compatible with control devices from a broad range of manufacturers, offering a high degree of choice as well as economy achieved through industry competition.

The wiring scheme adds labor and material costs to the installed system cost, but enables the dimming ballast to be linked to other ballast and control devices in a larger system, which in turn can be linked to local and centralized controls. 0-10VDC control allows the on/off control to be separated from the dimming control, allowing a combination of centralized switching and distributed dimming equipment to be used.

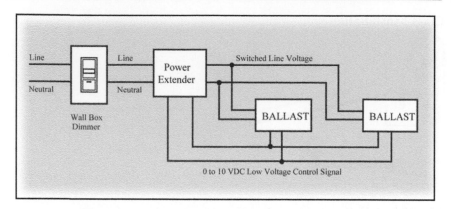

Figure 5-3. Schematic for a basic 0-10VDC dimming system. Graphic courtesy of Leviton.

0-10VDC ballasts are ideally suited for energy savings applications through a building management system and occupant control. In addition, 0-10VDC systems are required for the majority of applications that use photosensors. Another advantage of 0-10VDC control is the ability to control multiple lighting circuits from a single control location.

ANALOG TWO-WIRE PHASE-CONTROL METHOD

Analog two-wire phase-control is also called AC dimming, forward phase control, triac dimming, phase-chop dimming or two-wire dimming. It is a line-voltage dimming method that uses existing power wiring to communicate control signals between the controller and the ballast; both power and control are routed through the same line-voltage wires. As implied by its name, this ballast uses the same two line-voltage leads for both power and ballast control, wiring the same way as a conventional non-dimming ballast. The ballast receives the dimming signal through the dimmed hot wire connected to the power line.

The ballast reads the AC power supply signal's starting point or zero crossing point, then turns on the current after a preset waiting time. This cuts out part of the cycle and results in dimming. The waiting time, from 0 to 8.3 milliseconds or one-half the waveform, is related to the dimming level. See Figure 5-4.

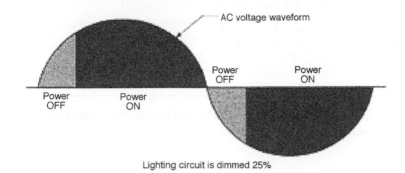

Lighting circuit is dimmed 25%

Figure 5-4. Analog two-wire phase-control dimming is a line-voltage dimming method that reads the AC power supply signal's starting point and then turns on the current after a preset waiting time, cutting out part of the cycle which in turn results in dimming. Source: Advance.

A dimming range of 100 percent to 5 percent is available for T8 lamps and CFLs, and 100 percent to 1 percent for T5HO lamps.

While two-wire ballasts can be incorporated into building-wide control systems, according to their primary manufacturer they are ideally suited for architectural dimming, stand-alone, retrofit and low-cost projects. New construction and retrofit applications include auditoriums and training areas, conference rooms and boardrooms, department and specialty stores, education, healthcare, hotels, houses of worship, private and executive offices, restaurants and other spaces. While the low-voltage wiring scheme of 0-10VDC systems lends itself to networked control, two-wire phase-control ballasts are typically controlled via local controls accessible to occupants.

Assessment

Two-wire phase-control ballasts have the advantage of not needing any additional wiring between the control device and the ballast, which makes them attractive for new centralized dimming

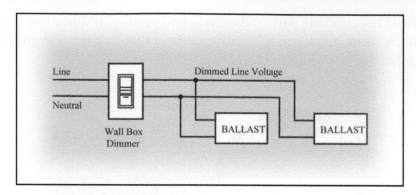

Figure 5-5. Schematic for a two-wire phase-control dimming system. Graphic courtesy of Leviton.

applications as well as retrofits due to a typically lower installed cost. They also don't require a separate switched power leg, so the hardware required to dim these ballasts is exactly the same as the hardware required to dim incandescent loads. This means that most (if not all) dimmer manufacturers include a way to adjust the dimming curve of their dimmers to allow the control of two-wire ballasts from their dimmer cabinets. It is stable over long wire runs, is considered easy to wire, and allows for maximum circuit loading.

Because the standard wiring configuration is utilized, phase-control dimming ballasts can represent a lower-cost dimming solution, typically found in architectural dimming applications such as conference rooms, boardrooms and individual offices. It is also ideally suited to retrofits of existing fluorescent fixtures, stand-alone applications and cost-sensitive projects. In addition, the control signals are less sensitive to interference than low-voltage analog signals.

Several manufacturers provide a selection of controls for two-wire phase-control ballasts, from stand-alone control solutions to larger dimming systems with dimming racks and/or cabinets. Integrating fluorescent dimming control with other load types on separately controlled zones can provide a system that is flexible and that can be tailored to the use of the space.

ANALOG THREE-WIRE PHASE-CONTROL METHOD

The three-wire phase-control configuration is based on the original magnetic dimming ballast conventions from the 1960s. This control method uses a third wire (in addition to hot and neutral) to carry the (typically) phase-control signal to the ballast. All three wires are rated Class 1 and can be run within the same conduit.

A dimming range of 100 percent to 1 percent is available.

Typical applications include conference rooms, boardrooms, patient/examination/treatment rooms, houses of worship, theaters, convention areas, restaurants, air traffic control centers, industrial control rooms, partitioned meeting rooms, graphic art workstations, CAD/CAM workstations and private offices.

Assessment

Three-wire phase-control ballasts draw very little current on the dimmed leg, which means that they can be dimmed without causing much heat to be generated at the dimmer. This allows devices that are intended only for this type of load to be smaller and also appropriate for use in a distributed system. It is stable over long wire runs, is considered easy to wire, and allows for maximum circuit loading.

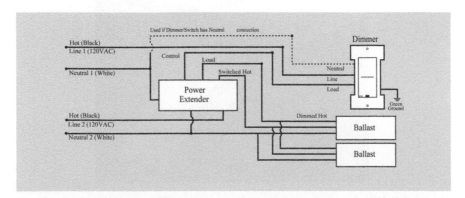

Figure 5-6. Schematic for a three-wire phase-control dimming system. Graphic courtesy of Leviton.

Three-wire ballasts are ideally suited as a building block of architectural dimming systems that can be integrated into building-wide systems.

ANALOG WIRELESS INFRARED

Wireless infrared dimming provides an individual control system that can also be integrated into a building-wide control system. This method uses an IR transmitter to perform the control function and does not require any additional wires (see Figure 5-7). The dimmer is included either in the ballast or as an additional device in the light fixture.

This can be a good retrofit solution, and allows for occupant control of their local lighting. Individual fixtures or groups of

Figure 5-7. Wireless infrared dimming uses an IR transmitter to perform the control function and does not require any additional wires. Graphic courtesy of Lutron Electronics.

fixtures can be controlled via a handheld remote control or PDA. In addition, these ballasts can be trimmed to reduce maximum light output.

A dimming range of 100 percent to 1 percent is available.

Typical applications include spaces where individual control is desired without additional wiring, such as conference rooms, board rooms, open and private offices.

Table 5-1. General comparison of analog dimming methods.

Analog Method	Available Dimming Range	Control Line/Low Voltage	Centralized/ Distributed	Manufacturers
0-10VDC	100 percent to 3 percent (T8 and CFLs) and 100 percent to 1 percent (T5HO)	Low	Centralized Distributed	Advance, Lutron, OSRAM Sylvania, Universal Lighting Technologies
Two-wire Phase-control	100 percent to 5 percent (T8 and CFLs) and 100 percent to 1 percent (T5HO)	Line	Centralized Distributed	Advance, Lutron
Three-wire Phase-control	100 percent to 1 percent	Line	Centralized Distributed	Lutron, Lightolier Controls
Wireless Infrared	100 percent to 1 percent	Wireless	Centralized Distributed	Lutron

DIGITAL METHOD (DALI)

Digital dimming can be used almost anywhere that analog dimming can be used, for the same purposes: visual needs, personal control, daylight harvesting, scheduling and other control strategies. If fluorescent dimming is desirable for a given application, digital dimming can offer distinct advantages related to intelligence, flexibility and two-way communication. It is particularly well suited for:

- Energy management applications such as scheduled automatic shutoff to meet energy codes, daylight harvesting—due to the ability to more economically set up very small control zones around the daylight aperture, and scheduling via a central computer, which allows load shedding, demand savings and potential utility incentives.

- Supermarkets, some retail spaces and similar applications with frequent merchandise or layout changes.

- Small and open offices where users are given dimming control over their own lighting as part of a strategy to increase worker satisfaction.

- Conference rooms, classrooms, training rooms and similar spaces that require different lighting scenes for multiple types of use.

- Larger installations with multiple buildings, where feedback on lighting component status can facilitate more efficient lighting maintenance.

Why DALI? In any dimming system, the ballasts and controllers must be able to "speak" and "hear/understand" the same language. In the case of digital dimming systems, this language is either proprietary—that is, unique to a particular manufacturer and, if allowed, other adopters—or an open standard: the Digital Addressable Lighting Interface (DALI) protocol. DALI, originally part of Europe's Standard 60929, has been a NEMA Standard (243-2004) in the United States since 2004.

When examining whether to use DALI for a fluorescent dimming installation, one must first assess the pros and cons of digital control, and then weigh the pros and cons of DALI as the communication protocol that enables the digital components to talk to each other.

Digital Dimming

The HVAC industry began embracing Direct Digital Controls (DDC) in the early 1990s. With digital electronic ballasts, this technology is now available for lighting. Digital dimming offers a number of clear advantages compared to analog dimming. These advantages are:

- Simplified wiring
- High degree of granularity of control accuracy (flexibility)
- Easy reconfiguration of control zones without rewiring
- Two-way communication (some digital ballasts)

In a digital dimming system, a single set of control wires form a low-voltage control bus—sometimes (inappropriately) called a loop.

Creating the Communications Network

Compatible ballasts and controls (up to a total of 64 devices—with each ballast having its own unique stored address) connect to this bus in order to provide control signal interaction. For larger installations, multiple buses can be networked to proper scale. The control bus provides two-way communication; ballasts can both respond to commands and reply to queries.

Control options include centralized systems (a personal computer or building automation system) as well as local controls such as manual dimmers, occupancy sensors and photosensors.

The ballasts and controls connected to the same bus can be assigned to up to 16 layers (groups or zones) of controls and scenes in the same space, and later reconfigured, via programming.

Simplified Wiring

A single pair of control wires, which form the bus, connect the ballasts and controls directly, which simplifies wiring in spaces with multiple control zones by reducing the number of homeruns (see Figure 5-9). The level of skilled labor is reduced because there is no need to pull wire according to a zoning schedule. Each ballast

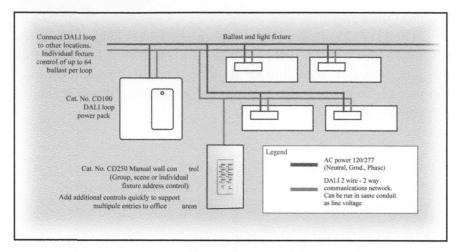

Figure 5-8. Schematic for a DALI-based digital dimming system. Graphic courtesy of Leviton.

on the circuit is wired the same. There are no switch legs or three-way travelers. In addition, dimming control panels/modules are not needed to control light output. Instead, digital systems use a small power supply connected to the bus.

Digital ballasts can be wired into the lighting system using Class 1 or Class 2 wiring methods in accordance with the National Electrical Code. Digital ballasts may use a Class 1-rated 5-conductor cable that uses one hot (live), one neutral, one ground and two polarity-insensitive control wires, all routed together in the same conduit. It is also possible to install the ballasts and controls as a Class 2 installation, in which case the control wires must be routed separate from the power wires. Check with the ballast and controls manufacturers whether their products are rated for Class 1 installation.

Flexibility

When configuring a fluorescent dimming system, the designer must specify control zones—a zone, in review, being a fixture or group of fixtures that is controlled simultaneously by a single controller. For example, in a daylight harvesting scenario with windows, the designer may place lighting circuits parallel to the

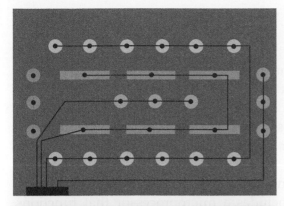

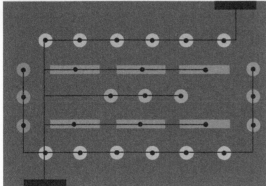

Figure 5-9. TOP: Conventional controls wiring scheme. Five separate lead runs from the wall controller are required in this example, creating wiring complexity. If the system must be changed in the future, rewiring is required. BOTTOM: Digital network wiring scheme. All ballasts are connected to the same lead runs, simplifying the wiring required. Rezoning and adding controllers are also simplified. Graphic courtesy of Universal Lighting Technologies.

window, and set up each circuit as a separate control zone. The smaller the control zones, the more granularity, or flexibility, can be achieved, and along with it higher energy savings. But cost also increases.

Digital lighting control provides the ultimate in flexibility. When using analog dimming systems, the smallest zone is a branch circuit. With a digital system, zoning is implemented using software, independent of circuits, using individual ballast addresses stored in memory. Because each ballast is individually addressable, control zones can be established that are as small as a single ballast or light fixture. Ballasts or fixtures can also be grouped to provide up to 16 layers of controls/scenes. This enables:

• Highly granular and responsive control

- Ability to generate a wide variety of zones and scenes in any controlled space

- Ability to adapt to changes over time and even be completely reconfigured without rewiring

For large installations, individual buses can be networked for the control of hundreds or even thousands of ballasts.

Two-way Communication

The digital ballast includes a microprocessor that functions as storage (ballast address, intensity settings, fade rate), receiver (control signals) and sender (intensity, lamp/ballast status) of digital information. DALI instructions such as GoToScene and SetMax are sent to the ballasts, utilizing the data stored in its microchip memory. But the ballast is not only "smart," it can also "talk back." As the control bus enables two-way communication, the digital ballast can not only receive commands, but respond with maintenance and energy information such as the status of the ballast and lamps.

This enables the lighting system operator to query ballasts for energy usage (using feedback such as IntensitySetting), which can be used for a variety of purposes from energy savings verification to benchmarking to billing internal departments or tenants individually for their lighting usage. It also enables the operator to query ballasts for lamp and ballast failure (querying for a response such as BadLamp), which can improve the efficiency of lighting maintenance and improve customer service from the facilities department.

Scheduled switching

With 0-10VDC dimming, the ballasts are connected with control wires; the controller can dim, but not switch, the lights. To achieve on/off switching, the power wiring would have to be reconnected to match the dimming circuits. With DALI-based digital lighting, any ballast or group of ballasts on the network can be given an on/off or DIM command without rewiring the circuits.

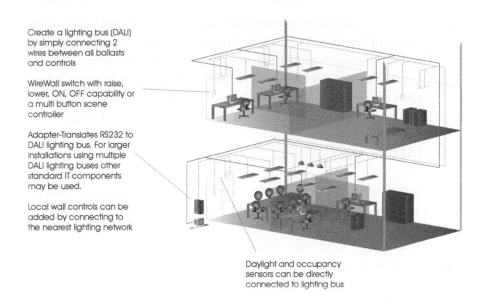

Create a lighting bus (DALI) by simply connecting 2 wires between all ballasts and controls

WireWall switch with raise, lower, ON, OFF capability or a multi button scene controller

Adapter-Translates RS232 to DALI lighting bus. For larger installations using multiple DALI lighting buses other standard IT components may be used.

Local wall controls can be added by connecting to the nearest lighting network

Daylight and occupancy sensors can be directly connected to lighting bus

Figure 5-10. For communication purposes, the lighting system uses a low-voltage control bus. The ballasts and all applicable controls, up to a total of 64 devices or 251mA, connect to the bus in order to provide control signal interaction. Graphic an amalgam based on image and information provided by Tridonic and OSRAM SYLVANIA.

This enables compliance with prevailing energy code requirements for automatic shutoff in controlled spaces without the need for a control panel with a scheduling function.

Buyer Beware

The higher level of capabilities from digital control systems often entails tradeoffs in cost and complexity.

Digital ballasts and controls typically present a higher component cost largely due to power supply/router requirements, but they can present a lower installed cost due to a reduction in wiring labor for group and scene control, and the removal of the need for dimmer modules/control panels in larger installations.

Digital control also presents more sophisticated programming when centralized systems are used, and requires on-site field commissioning, which should be factored into the design specification so that the appropriate party is aware and can bid on this portion of the installation. During the start-up phase, a database of ballast addresses, with physical locations and the control device that operates them, needs to be created and then maintained when the layout of the space changes. It may be advisable to involve the client's IT staff in the creation and maintenance of the database.

The DALI Protocol

Once the designer decides to implement a digital control strategy, a primary choice is whether these devices should communicate using a proprietary protocol or DALI, which is a standard open protocol.

Proprietary protocols: The advantage of a proprietary protocol is that the complete control system can be furnished by a single manufacturer, which has tested all components to ensure interoperability and supports the entire system on the job. The disadvantage is that the owner is tied to a single manufacturer, which limits choice and potentially sacrifices economy.

DALI protocol: DALI is a royalty-free, non-proprietary, two-way, open and interoperable digital protocol. Currently, DALI consists of a set of commands to and from ballasts within a defined data structure and specified electrical characteristics. As an open standard (NEMA 243-2004), DALI is supported by five major U.S. electronic ballast manufacturers and a growing number of controls manufacturers, which offer DALI-compatible products.

The advantages of DALI are that it:

- Provides true interchangeability across ballasts and controls. Multiple manufacturers can be involved in building an appropriate solution, instead of being tied to a single supplier, using DALI as an open platform. This can result in lower costs, ensures future availability, and enables the system designer to select product functions from one manufacturer and combine them with products from other manufacturers.

- Provides standardized ballast performance. For example, DALI defines light output for all levels of dimming signals. DALI ensures consistent dimming performance across all dimming ballasts regardless of type or manufacturer, currently not achievable with analog dimming methods such as 0-10VDC.

Ballasts and Controls

DALI-compatible fluorescent dimmable electronic ballasts are currently available in:

* Universal input voltage (120VAC to 277VAC)
* One-, two-, three- and four-lamp models for T8 lamps
* One- and two-lamp models that operate CFL, T5 and T5HO lamps

Ballasts: Digital ballasts are available that provide a dimming range of 100 percent to 1 percent, utilizing a logarithmic dimming curve. The inverse-square dimming curve is used for better control of the lighting intensity in response to the human perception of brightness.

Digital ballasts utilize programmed-start technology to maximize lamp service life, highly suitable for frequently switched applications such as installations with occupancy sensors.

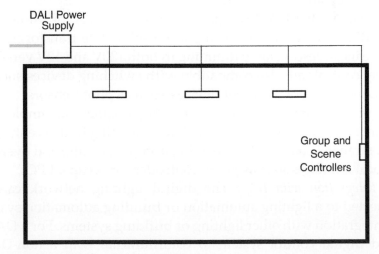

Figure 5-11a: Stand-alone DALI application. Courtesy of Lightolier.

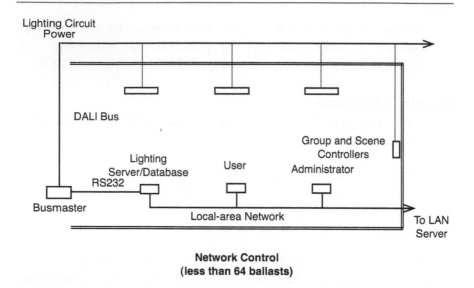

Lighting Circuit
Power

DALI Bus

Group and Scene
Controllers

Lighting
Server/Database
RS232

User

Administrator

Busmaster

Local-area Network

To LAN
Server

Network Control
(less than 64 ballasts)

Figure 5-11b: Small networked DALI application. Courtesy of Lightolier.

Controls: DALI based digital control systems (see Figure 5-12 for examples) can include all controls that would normally be used for multi-scene fluorescent dimming, such as preset controls, as long as they are rated as compatible with DALI. They can be used to operate with a DALI compatible digital ballasts for fluorescent lamp dimming, or a DALI compatible solid-state transformer for precise incandescent lamp dimming in both 120V and 12V versions. Digital controls are also compatible with switching devices such as occupancy sensors and other devices such as photosensors.

With a digital control scheme, the designer can implement automation strategies such as scheduled dimming from a central PC for centralized control while enabling occupant control and override via local interfaces such as preset controllers or occupant PCs.

Integration with BAS: The digital lighting network can be connected to a lighting automation or building automation system for integration with other lighting or building systems. For a DALI-based digital control system to communicate with a non-DALI lighting control system, or a BAS operating using another protocol

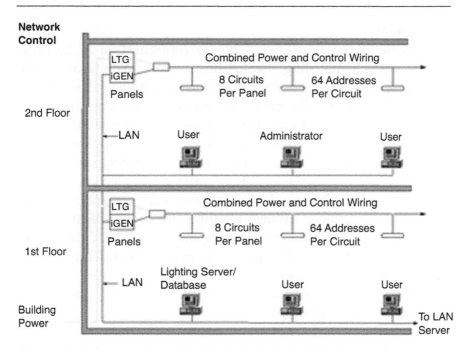

Figure 5-11c: Large networked DALI application. Courtesy of Lightolier.

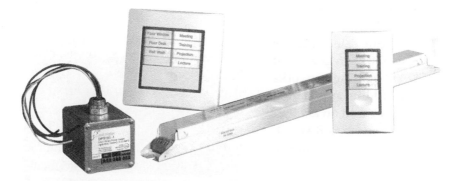

Figure 5-12. DALI-compatible digital lighting controls. Photo courtesy of Watt Stopper/Legrand.

such as BACnet or LonWorks, a translator device, called a gateway, is required to enable these systems to communicate with each other (see Figure 5-13).

Assessment

Digital dimming allows software configuration of lighting groups, presets matching the lighting to the space usage, and integrated energy management functions. Digital systems can be configured as large networked systems requiring commissioning and training, or as simple stand-alone room preset dimming controls requiring no special tools or PCs. As an open standard used with digital systems, DALI enables true interchangeability among vendor products and standardized performance across manufacturers. Although digital systems can present a higher component cost, labor savings resulting from simplified wiring can result in a lower installed cost compared to 0-10VDC dimming. And although digital dimming is new, and a little different, much of the equipment and methods will be familiar to designers and installers of 0-10VDC dimming systems. In several important ways, installation is actually simpler.

Digital dimming is not for all applications, but it offers clear advantages in applications where fluorescent dimming is both well-suited and desired. As demand for fluorescent dimming increases, so will the demand for digital dimming increase as a control method that offers distinct benefits.

Figure 5-13. Gateway enabling integration of DALI lighting networks with building automation systems based on another protocol such as BACnet. Photo courtesy of Square D.

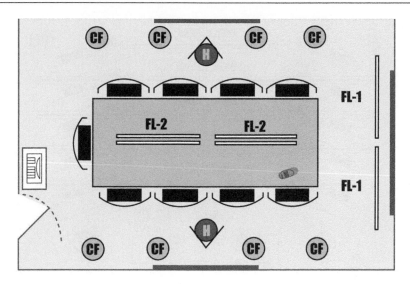

Figure 5-14. Digital system in a conference room. Components include:

(2) 1xF32T8 dimming ballasts (FL-1)
(2) 2xF28T8 dimming ballasts (FL-2)
(8) 1xCFQ26W dimming ballasts (CF)
(2) electronic halogen dimming modules (H)
(1) wall-mounted 4-scene recall controller (SC)
(1) handheld programmable remote (HH)
(1) power supply (PS)

In this space, light levels can be controlled so that the side of the room where presentations occur can receive more or less illumination depending on the media used. Halogen accent lights highlight pictures on two of the walls. The 4-scene wall-mounted controller is used at the entrance. The handheld remote is used for zone programming as well as programming scenes and recall. With 14 light sources in this room, most of them can be established as its own zone for a very high degree of flexibility and control resolution. Graphic courtesy of Universal Lighting Technologies.

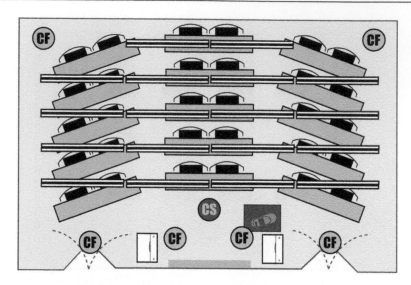

Figure 5-15. Digital system in a lecture hall. Components include:

(18) 2xF32T8 dimming ballasts
(6) 1xCFM32W dimming ballasts
(1) ceiling-mounted IR sensor kit
(2) wall-mounted on/off, up/down controls
(1) handheld programmable remote control
(1) handheld remote with 4-scene recall
(1) power supply

In this space, control zones can be established so that during A/V presentations, the lights near the front of the hall can be dimmed while the lights in the rear can be set at higher levels. A 4-scene handheld remote is used with the ceiling sensor for dimming control. Controls at the doors provide both on/off and up/down control. The programmable remote is used for programming and is then stored only with access to authorized users. Graphic courtesy of Universal Lighting Technologies.

Table 5-2. Detailed comparison of dimming methods.

Method	0-10VDC	Two-Wire Phase-Control	Three-Wire Phase-Control	Digital (DALI)
Location	Typically found in applications where energy management is the primary goal, such as open offices, etc.	Typically found in architectural dimming applications such as conference rooms, cove lighting, private offices, etc.	Typically found in architectural dimming applications such as conference rooms, cove lighting, private offices, etc.	Found in both energy and architectural applications
Wiring Needs	Four-wire installation, with two wires for powering the ballast and two for changing the light level; requires a network of low-voltage wiring to tie the ballast together into a system.	Two-wire installation; wires are used for both powering the ballast and changing the light level; ballasts easily connect to existing line-voltage wires.	Three-wire installation; wires are used for both powering the ballast and changing the light level.	Four-wire installation with two wires for power and two for data, all run in the same conduit.
Uses	Due to the wiring configuration of the ballast, the owner can tie the lighting system together with occupancy sensors, daylight sensors, building management systems and manual dimming controls.	Due to easy installation and stand-alone operation, these ballasts are highly suitable for spot dimming projects, retrofits, architectural dimming applications, and in projects where a low installation cost alternative and/or a faster payback is required.	Architectural dimming for conference rooms and similar applications where a wider dimming range is desired.	Primarily used for new construction with all of the features of 0-10VDC plus the ability to control individual fixtures in open office applications. Also provides central reporting of lamp or ballast failures.

Table 5-2. Detailed comparison of dimming methods (*Continued*).

Method	0-10VDC	Two-Wire Phase-Control	Three-Wire Phase-Control	Digital (DALI)
Operating controls	Energy management operators typically automatically operate the dimming system based on a schedule or local input such as ambient light for daylight harvesting; total lighting control.	Individuals typically manually operate the dimming system based on their light level preferences.	Individuals typically manually operate the dimming system based on their light level preferences.	Supports both individual control and central automation.
Cost	Higher installed cost, but higher energy savings.	Lower installed cost, and good quality of dimming.	Higher ballast cost, but excellent quality of dimming.	Highest ballast cost but lowest installed cost for individual fixture control, which provides maximum energy savings and occupant satisfaction.
Availability	Can use ballasts from most major ballast manufacturers and most controls manufacturers.	Can use ballasts from Advance and controls from most major controls manufacturers.	Can use ballasts from Lutron and Lightolier Controls, and controls from the same manufacturers and a limited number of other manufacturers.	Ballasts available from four major manufacturers, controls from major controls suppliers.

Chapter 6

Lamp/Ballast Interactions

Dimmable ballasts work with lamps as a system, resulting in interactions between these components that affect lamp performance. While lamps operated by dimmable magnetic ballasts had experienced problems such as flicker, today's dimmable electronic ballast-operated systems generally do not experience these problems.

In this chapter, you will learn about lamp life; flicker, flashing and other visual instabilities; premature lamp end blackening; color shift; remote ballast mounting; and harmonics as these factors relate to dimmable ballasted systems.

DIMMING AND LAMP LIFE

Whether dimming or fixed output, optimizing lamp life depends largely on the ballast's ability to provide sufficient cathode heating at startup and during operation.

Dimmable magnetic ballasts operating linear lamps at low levels of light output for long periods of time can shorten lamp life. However, this does not appear to be true for dimmable electronic ballasts *when installed and operated properly*.

One might think that dimming increases fluorescent lamp life, but this is only true for incandescent lamps. Comprehensive life-testing of dimmable electronic ballasts indicates that prolonged dimming, when operated properly, neither decreases nor increases fluorescent lamp life. Some manufacturers clearly indicate in their performance specifications that operating the given ballast will not affect average rated lamp life. (*However, this issue continues to be the subject of research.*)

Prolonging Lamp Life

Dimming ballasts that are programmed-start ballasts may increase lamp life because of reduced wear and tear on the lamp during startup.

Some ballasts can only dim to a specific level and must turn on at a specific level. Turning it on or dimming below this level can result in premature end of lamp blackening, lamp failure and performance issues such as flickering. For more information, consult with the ballast manufacturer's performance specs for the ballast model of interest.

Newly installed dimmable linear and compact fluorescent lamps should be operated at full light output for a recommended period of time prior to any dimming.

Lamp life can be reduced if the lamp filaments do not make proper contact in the socket. Be sure to use only rapid-start sockets, with knife-edge sockets being recommended rather than flat-edge sockets.

Lamps can experience improper starting and shortened life if the leads between the ballast and the lamps exceed the maximum length published by the manufacturer. Consult the ballast manufacturer regarding the maximum lead length for your lamp type and ballast.

FLICKER, FLASHING AND VISUAL INSTABILITIES

Properly specified, installed and operated, dimmable electronic ballasts do not cause lamp flicker, flashing or other visual instabilities.

Flicker

Magnetic ballasts, which operate at 60Hz frequency, produce some flicker that is generally not noticeable but may cause eyestrain and headache among some occupants. Electronic ballasts, which operate at 20,000+ Hz frequency, reduce lamp flicker to an essentially imperceptible level due to high-frequency operation.

Lamp Seasoning

Newly installed fluorescent lamps may flicker when operated by any type of ballast because of residual impurities from the manufacturing process, or because of how mercury is distributed in the lamp. Operating the lamps at full light output for a period of time, called burn-in or lamp seasoning, is often all that is needed to iron out these causes of flicker and other visual instabilities.

NEMA's Lighting Systems Division addressed the issue by creating a new recommended practice in 2002. NEMA fluorescent lamp manufacturers recommend that whenever flicker or instability occurs, or to avoid it when new dimming systems are commissioned, the lamp should be operated at full light output overnight (about 12 hours). Overnight "seasoning" is particularly recommended, says NEMA, for optimal initial performance in installations where dimming performance (tracking, stability) is considered critical.

Another point of view can be found in the 2001 *Advanced Lighting Guidelines* published by the New Building Institute, which recommends a 100-hour burn-in period at full light output for all fluorescent lamps prior to dimming.

For lamps that require seasoning, group relamping is sometimes recommended to maintain uniform dimming performance from the entire system. Group relamping entails replacing all lamps in the lighting system at periodic intervals, usually 60-70 percent of rated life, when the mortality rate for linear fluorescent lamps begins to escalate. Group relamping is a sound strategy for any fluorescent lighting system because it can improve light levels and space appearance while economizing on maintenance labor costs.

In recent years, however, a number of lamp manufacturers have begun to take the view that most of their linear fluorescent lamps can be dimmed immediately, while dimmable compact fluorescent lamps require a 100-hour burn-in at full light output (about four days) prior to dimming. Failing to allow a sufficient burn-in period for newly installed dimmable compact fluorescents can result in striations, severe lamp-end blackening, poor dimming performance, and premature lamp failure. For the latest information, consult the lamp manufacturer.

Other Causes of Flicker

Flicker in linear fluorescent and compact fluorescent lamps operated by dimmable electronic ballasts is not always due to the manufacturing process. Below are six possible causes:

If flicker persists after a suitable burn-in period, the lamp itself may be defective. You may try replacing the lamp, operate it for a sufficient burn-in period, and then determine if the problem is with the lamp or with the system.

Some dimmable ballasts can only be dimmed to a specific level of light output and must start at a minimum light output. Turning it on or dimming below this level can result in lamp flicker at the low end of the dimming. Consult the ballast manufacturer for the performance specifications of the particular ballast being used, and be sure to operate it within those specs. Another solution for installed ballasts is to raise the low-end light output.

Electrical noise on the line can also cause lamp flicker at all dimming levels.

Improper wiring and or installation can cause flicker. Ensure that wiring from the dimmer to the ballast, and the wiring from the ballast to the sockets, is done according to manufacturer specifications. Also ensure a proper maintained connection to ground.

Dimmer-ballast incompatibility can cause flicker—ensure that the dimmer and the ballast are compatible.

Fluorescent lamp sockets may feature mounting slots to allow variation in the height of the lamp from the grounded metal surface. These slots enable the installer to position the outside edge of the lamp an appropriate distance from the grounded metal surface. Mounting the lamp too far from the grounded metal in the fixture may result in lamp flicker or inability to light. Mounting the lamp too close to the grounded metal may result in premature lamp failure and poor intensity at the low end of the dimming range.

Flashing

If flashing occurs, often it is due to incorrect wiring or incompatibility of components. One ballast manufacturer says that

90 percent of its field service calls to address flashing and visual instability in lamps in electronically dimmed fixtures are the result of incorrect wiring. Installers should take special care to wire the system in accordance with the manufacturer-supplied wiring diagram.

Flashing and instabilities may also occur when components in the system are not compatible with each other. For example, the ballast could be wired to the wrong type of lamp sockets. Be sure to use only rapid-start lamp sockets, with knife-edge sockets being recommended rather than flat-edge sockets to ensure good filament contact.

Lamp End Blackening

Lamp end blackening occurs naturally as the lamp approaches failure. Premature lamp end blackening should not occur in dimming installations that are specified, installed and operated properly.

If premature lamp end blackening occurs in a dimming installation, it may be caused by lamp-ballast incompatibility, faulty connections, faulty ground connections, or a defective lamp.

COLOR

Linear lamps operated by dimmable magnetic ballasts may experience color shift at low levels of light output.

Dimmable electronic ballasts may cause lamps to experience some color shift towards the blue or cooler part of the spectrum (incandescent lamps shift towards the yellow or warmer part of the spectrum) as the lamps are dimmed. The color temperature (K) of the fluorescent lamp increases slightly as fluorescent lamps are dimmed, as opposed to decreases, as with incandescent lamps.

DIMMING COLD LAMPS

Ambient temperature can affect the startup and operation of fluorescent lamps. Optimal dimming occurs when the lamps are warm and operated at the manufacturer specification (typically 77°F). If the lamps are cold, startup and operation at both the low end and the high end of the dimming range may be affected.

When lamps are operated at 100 percent for several minutes, they reach their normal operating temperature. When dimmed to absolute low, they will typically dim down to close to their specified dimming range. As the lamps remain at this low level, they will cool and continue to dim down slightly. This should not be noticeable to the occupant of the space. The same is true when going from prolonged use at absolute low to 100 percent, as the lamps slightly increase lumen output as they warm up.

BALLAST LIFE

The service life of the dimmable electronic ballast is similar to that for non-dimmable rapid-start and programmed-start electronic ballasts operating linear lamps. However, as with any ballast, service life is largely dependent on ballast case temperatures experienced in the field. (Ballasts generate heat and must be able to dissipate it to avoid premature failure.)

If the maximum ballast case temperature (70°C or 158°F) is not reached, average rated ballast life is often listed at 60,000 hours. If this threshold is crossed, ballast life will be reduced and the ballast's warranty may become affected. Another major source of ballast failure is from incorrect wiring and poor connections.

For integral dimmable electronic CFLs, the average rated ballast life is limited to the life of the lamp. For non-integrated systems, average rated life is often listed at 50,000 hours for ballasts operating lamps as a two-piece system.

For every 10°C increase in ballast case temperature, ballast life can be expected to be reduced by one-half. Conversely, reducing ballast case temperature will increase ballast life. For every 10°C reduction in ballast case temperature, ballast life can be expected to double.

Note that improper ballast installation can affect the device's ability to dissipate heat. For best results, ballasts must be mounted flush to the fixture, and not mounted on the fixture cover plate that holds the lamps, as this is often the hottest spot in the fixture. Screws, knockouts or any other feature that raises some or all of the ballast off the fixture are not acceptable, and these will affect the ballast's ability to transfer heat.

REMOTE BALLAST MOUNTING

Generally, dimmable electronic ballasts designed to operate three or four linear lamps must be operated in the same fixture as the lamps. The same holds true for compact fluorescent ballasts.

Dimmable electronic ballasts designed to operate one or two linear lamps can be remote-mounted or tandem-wired, as long as the ballast is rated for this. The maximum allowable remote-mounting distance may vary from one type of ballast to the next. Identical installations with different fixtures may see variable results due to the internal wiring path of the fixture. For this reason, it is important to mock up atypical applications. Consult the ballast and fixture manufacturers.

HARMONICS

Total harmonic distortion (THD) is the ratio of 1) the sum of all harmonic frequencies above the fundamental frequency to 2) the value of the fundamental frequency. High levels of THD present in an electrical system can damage equipment and pose a fire hazard.

THD became an issue in the early 1990s due to the rapid adoption of electronic ballasts. The ballast manufacturers responded by using passive filters to limit THD to less than 10-20 percent. Electronic ballasts rated <10 percent THD (and harmonic-canceling transformers) can be specified for buildings that have strict power quality requirements, such as hospitals.

Harmonics and Low-Voltage Dimming Systems (0-10VDC)

THD has been reported by the Lighting Research Center to increase on 0-10VDC dimmable ballasts as light output decreased. The increase in THD in turn decreased power factor. The Lighting Research Center concluded that since THD is a percentage of the fundamental current, a high THD at low fundamental current levels associated with low light output levels may not be a concern, as the actual distorted current is small. However, THD is still about 20 percent at full dimming with 0-10VDC dimmable ballasts.

Harmonics and Line-Voltage Dimming Systems (Phase-Control)

Because THD increases during dimming, one may be concerned about stressing the power line neutral, particularly in phase-control dimming systems that use line-voltage power wires for distribution of control signals. Since the input current to the ballast decreases as it dims the lamp, however, THD in the neutral remains relatively constant throughout the dimming range. Therefore, if THD is not presently an issue in the environment, converting to dimming will not have an adverse affect. To further address concern about overloading the neutral, dimmable electronic ballasts can be specified with <10 percent THD (rated at full light output).

EFFICACY

The efficacy of instant-start T8 systems is about 5 percent greater than rapid-start T8 systems (except for new high-efficiency ballasts) and 9 percent greater than dimming rapid-start systems (see Table 6-1). Energy savings gained during dimming periods can be eroded during non-dimming periods relative to instant-start systems. New dimmable ballasts are being introduced by at least one manufacturer, however, that approximately match the efficiencies of standard instant-start ballasts.

Proper operation of dimming ballasts requires programmed-start rapid-start operation, which maximizes lamp life at the cost of additional heating to the cathodes and subsequent system energy consumption. For a given 0-10VDC ballast at 3 percent lamp output, the system consumes 19 percent of full input wattage. For a given phase-control ballast at 5 percent lamp output, the system consumes 22 percent of the full input wattage. This demonstrates that useful energy savings are not achieved below about 20 percent dimming.

Table 6-1. General comparison of instant-start ballasts and typical programmed-start dimmable ballasts.

Lamps	Voltage	Starting	Interface	Ballast Factor Max.	Min. Dimming Range	ANSI System Watts Max.	Min. Watts
(2) F32T8	120V	Instant start	Fixed light output	0.88	NA	59	NA
(2) F32T8	120V	Instant start	Fixed light output	1.01	NA	65	NA
(2) F32T8	120V	Programmed start	0-10VDC	1.00	3%	68	13
(2) F32T8	120V	Programmed start	2-wire Phase-control	1.00	5%	68	15

SPECIFICATION TIPS

Use the Right Sockets: Rapid-start lamps and sockets are recommended for dimming. The socket should secure the lamp in place to ensure proper lamp contact, which is critical for proper dimming. Also note that if the ballast operates more than one linear lamp, the lamps may need to be wired in parallel and not in series. See Table 6-2.

Table 6-2. Lamp sockets recommended for dimming.

Lamp Type	Recommended Sockets
T8 and T12	Rapid-start knife-edge/Rotary locking
T5 linear	Rotary locking
T4 CFL	4-pin
T5 twin-tube	Socket that "locks" lamp in place

Avoid Mixing Loads on Same Dimmer: It is recommended to avoid mixing different loads on the same dimmer, such as incandescent lamps and fluorescent ballasts. With dimming systems, different loads can be mixed on the same control station, each with their own zone, with the use of a different dimmer at the panel designated for each load.

Operate the Ballast within Parameters: Note that some devices can be dimmed only to a specific level. While some ballasts may provide lamp ignition at any level, some may also require a minimum turn-on level. Ensure that the ballast is not dimmed or turned on below these levels. Some manufacturers provide a low-end trim feature to avoid this.

Account for Voltage Leak: Many dimming systems leak voltage to the ballast when "off," which can cause the ballast to fail prematurely and also cause lamp flicker. It is recommended that the dimmer be positively disconnected from power when it is off. This may be accomplished with a relay in line. Consult the controls manufacturer.

Account for De-rating: The current-carrying capacity of dimmers may be expressed in amps, watts, ballasts, etc. and may be also based on the voltage of the AC power supply. It is recommended that circuits feeding ballasts should not exceed 80 percent capacity and should be de-rated accordingly. Because of large in-rush currents, magnetic ballasted loads should be de-rated up to 50 percent. De-rating may also be required for ganging dimmers.

Account for In-rush Current: When most lighting loads are

switched on, they can draw a much higher level steady-state current for the first couple of line power cycles—a fraction of a second. In some electronic ballasts, particularly ones using front-end active filters to reduce harmonic distortion, this in-rush current may cause the contacts in lighting relays to fuse and damage connected devices such as dimmers. It is recommended to size any lighting relays to account for this. In addition, some dimmable ballasts feature in-rush current-limiting circuitry. NEMA Guideline 410-2004 sets voluntary guidelines for manufacturers of lighting controls and switching devices with electronic ballasts; look for ballasts that meet these guidelines.

Ensure Compatibility and Interoperability: Be sure that the selected ballast is dimmable, that the ballast is rated to operate the selected lamp, and that the ballast and controls are compatible. To ensure consistent dimming performance, it is recommended that all lamps controlled by a dimmer should be of the same brand and wattage. (Note, however, that many current dimmable ballast designs allow for similar operation of 17W, 25W and 32W lamps on the same circuit, typically in cove lighting applications.) Similarly, all ballasts controlled by a dimmer should of the brand and model.

Account for Power Draw in off State: Some dimmable ballasts are designed to switch the lights completely off after reaching the low end of their dimming range, but will continue to draw power while off. This should be accounted for in energy calculations. Consult the manufacturer.

Stay within Maximum Lead Lengths: The ballast manufacturer will provide a maximum length for the leads connecting the ballast and the lamp. Exceeding these maximum lengths can result in poor dimming performance and early lamp failure.

Stay within Maximum Control Wire Lengths: If the dimming system is based on the 0-10VDC dimming method, the control manufacturer will provide a maximum length for the control wires running between the dimmer and the ballast. To prevent dimming performance being affected by electrical noise or interference, consider conduit or shielded cable if the wiring run length must be extremely long and/or electrical noise or interference is present.

This may also be a code requirement in some regions. Note that digital ballasts also use low-voltage control wiring but is not subject to noise or interference.

INSTALLATION TIPS

Grounding: The lighting fixture, ballast and dimmer must be grounded in accordance with safety considerations and the National Electrical Code. Grounding the ballast to the fixture requires star-shaped screws, washers or nuts in order to penetrate the paint finish on the ballast. Both ends of the ballast must be attached to the fixture to ensure proper grounding (see Figure 6-1).

Ballasts generate heat and must be able to dissipate it to avoid premature ballast failure. For best results, ballasts must be mounted flush to the fixture, and not mounted on the fixture cover plate

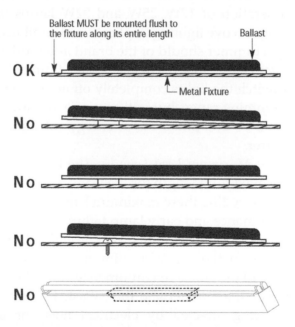

Figure 6-1. Grounding the ballast. Graphic courtesy of Lutron Electronics.

that holds the lamps, as this is often the hottest spot in the fixture. Screws, knockouts or any other feature that raises some or all of the ballast off the fixture are not acceptable, and these will affect the ballast's ability to transfer heat.

Lamp Mounting Height: Fluorescent lamp sockets may feature mounting slots to allow variation in the height of the lamp from the grounded metal surface. These slots enable the installer to position the outside edge of the lamp an appropriate distance from the grounded metal surface. Mounting the lamp too close the grounded metal may result in premature lamp failure and poor intensity at the low end of the dimming range. Mounting too far may result in lamp flicker or inability to light. See Figure 6-2.

Testing Dimmers

Control Test (0-10VDC): Disconnect power by turning off the breaker. Disconnect the dimmer from the power and ballasts. Use a true RMS digital multi-meter, designed for use with high-frequency operation, to measure the resistance from the purple to the gray leads; the resistance should change as the dimmer is adjusted.

Control Test (Two-Wire Phase-Control): Disconnect power by turning off the breaker. Disconnect the dimmer from the ballast. To see, in general, if the dimmer is operating, connect a 40W

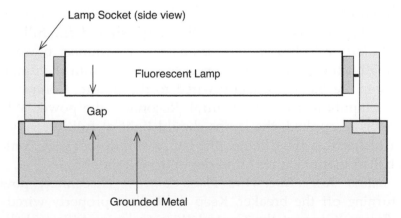

Figure 6-2. Lamp mounting height. Graphic courtesy of Lutron Electronics.

incandescent lamp to the dimmer's output leads and then reconnect the power; when power is re-applied, the lamp's intensity should change as the dimmer is adjusted. For a more detailed analysis of the dimmer's operation, use a true RMS digital multi-meter, designed for use with high-frequency operation, to measure the output voltage of the dimmer. Confirm this measurement meets the manufacturer's requirements.

   Control Test (Three-Wire Phase-Control): Disconnect power by turning off the breaker. Connect a true RMS digital multi-meter from the Control Wire (color specific to dimmer manufacturer) to Neutral (white) and set the meter to measure DC voltage. Reconnect the power. When power is re-applied, the voltage should change as the dimmer is adjusted from high-end to low-end.

   Confirm the most appropriate testing method with the control manufacturers.

Testing Ballasts

   Ballast Test (0-10VDC): Disconnect power by turning off the breaker. Disconnect the purple and gray wires from the dimmer but keep the ballast properly wired to the fixture and power. Reconnect the power. When power is re-applied, the lamps should light at full intensity. To check if the ballast is dimming properly: With power to the ballast, touch the low-voltage leads (gray and violet) of the ballast together and the lamps should dim fully.

   Ballast Test (Two-Wire Phase-Control): Disconnect power by turning off the breaker. Disconnect the ballast from the dimmer but keep the ballast properly wired to the fixture. Connect the ballast inputs to Hot and Neutral. Reconnect the power. When power is re-applied, the lamps should light at full intensity. If, after reattaching the dimmer, the lamps still do not dim, contact the ballast manufacturer for in-depth troubleshooting.

   Ballast Test (Three-Wire Phase-Control): Disconnect power by turning off the breaker. Keep the ballast properly wired to the fixture. Connect the Control Wire (color specific to ballast manufacturer) to ground. Reconnect the power. When power is re-applied, the lamps should dim to absolute low intensity.

Confirm the most appropriate testing method witih the ballast manufacturer.

Table 6-3. Troubleshooting fluorescent dimming systems.

Problem	Possible Causes	Prescription
Premature lamp failure	Miswiring	Verify fixture wiring versus wiring diagram on ballast. Pay special attention to "shared" wires
	Incorrect lamp type	Use only rapid-start lamps
	Lamp not making good contact in socket	Make sure proper socket is used and that lamp is making good contact
	Lamp and ballast are not compatible	Verify compatibility
	Ballast not properly grounded	Check ballast mounting and wiring. Also make sure the fixture is properly grounded
	Lamp dimmed too low	Raise low-end setting
	Ballast-to-lamp lead lengths exceed specified maximum	Verify lead length is within maximum
	Defective lamp	Replace lamp
Premature ballast failure	Poor heat dissipation	Check wiring diagram
		Check ambient heat in area
		Ensure ballast is properly mounted and grounded
	Control is leaking voltage to ballast in off state	Use control that provides positive disconnection from power when off
	Ballast is dimmed to lower level than it can handle	Raise low-end setting
Premature dimmer failure	In-rush current from ballast	Use ballast that has in-rush current-limiting circuitry. Confirm compliance with NEMA Guideline 410-2004
		Use control that is designed to handle in-rush current
	Circuit is overloaded	Size dimmer appropriately, add power pack, or reduce load on circuit

(Continued)

Problem	Possible Causes	Prescription
Lamp flicker	Poor connections	Check connections at sockets, connect-orsat ballast, wire nuts. Look for pinched wires causing intermittent shorts
	Ballast is dimmed to lower level than it can handle	Raise low-end setting
	Line noise	Try on different breaker
		0-10VDC: Control wires should not exceed maximum wire run length
		0-10VDC: Consider conduit or shielded cable
		Add choke or inductor to line
	Ballast-to-lamp lead lengths exceed specified maximum	Verify lead length is within maximum
	Defective lamp	Replace lamp
Lamps do not dim	Control wires reversed	Verify proper connections and orientation of control and power wires (0-10VDC ballast)
		Verify proper connections and for Hot and Dimmed Hot wires (three-wire phase-control ballast)
	Dimmer not communicating with ballast	Verify dimmer status and wiring to ballast
		Verify compatibility of ballast and control
		Ensure that line and 0-10V low-voltage wires are separated by conduit
	Processor failure at dimming panel (system dimmer with distributed intelligence)	Check the panel. Check address set-up
	Other problem at panel	Check the panel
Lamps do not sufficiently dim	Ballast's dimming range	Check ballast performance parameters
		Check ground
	Miswiring	Verify fixture wiring versus wiring diagram on ballast. Pay special attention to "shared" wires

(Continued)

Problem	Possible Causes	Prescription
Lamps have problems starting or do not start	Incorrect lamp type	Use only rapid-start lamps
	Lamp not making good contact in socket	Make sure proper socket is used and that lamp is making good contact
	Lamp and ballast are not compatible	Verify compatibility
	Ballast-to-lamp lead lengths exceed specified maximum	Verify lead length is within maximum
	Defective lamp	Replace lamp
Lamps cannot be turned off	Circuit is overloaded	Size dimmer appropriately, add power pack, or reduce load on circuit; check that dimmer has not been shorted
Inconsistent dimming performance	Brand-new lamps	Operate lamps at full light output for period specified by lamp manufacturer prior to dimming
	Different lamp and ballast brands mixed on same dimmer	Use only one lamp brand and wattage, and one ballast brand and model, on a single dimmer
Premature lamp end blackening	Miswiring	Verify fixture wiring versus wiring diagram on ballast. Pay special attention to "shared" wires
	Lamp and ballast are not compatible	Verify compatibility
	Lamp not making good contact in socket	Make sure proper socket is used and that lamp is making good contact
	Ballast not properly grounded	Check ballast mounting and wiring
	Defective lamp	Replace lamp
Lamp undergoes slight color shift towards blue during low-end dimming	Normal for fluorescent, incandescent and HID to undergo some color shift during dimming	Raise the low-end setting; insufficient data exist about this issue, however, for manufacturers to be able to recommend how much to raise it
Audible noise	Panel noise	Use panel that dampens fan noise or use a convection-cooled panel
		Isolate the panel from acoustically sensitive spaces
	Ballast is dimmed to lower level than it can handle	Raise low-end setting

Chapter 7

Dimming Controls

Dimming controls can be categorized as basic wall-box dimmers, integrated dimmers, modular dimmers, low-voltage dimmers and preset dimmers. Dimming controls are available for most types of lighting, including fluorescent systems. They can offer manual or automated control or a combination of the two. As a strategy, dimming enables flexible control of lamp output and power to save energy and/or support visual needs/tasks.

While large, sophisticated dimming systems are increasingly being deployed as an energy management strategy, many dimming systems are relatively simple layouts serving a single room or space. The ballasts can be commanded by connected devices such as occupancy sensors, which signal the ballasts to switch on/off based on occupancy, and photosensors, which signal the ballasts to dim when light levels are sufficiently supplemented by available daylight. Often, however, the ballasts are commanded by a controller called a wall-box dimmer.

WALL-BOX DIMMERS

In this section, you will learn about the function of the wall-box dimmer, typical applications, how they work, dimming methods, types of dimmers, configurations for controlling a lighting system from single or multiple locations, ganging dimmers in the same location for control of multiple lighting circuits, sizing the dimmer for the application, power packs to increase the dimmer's capacity, and compatibility issues.

Wall-box dimmers are often applied in premium spaces such as executive offices and multi-purpose rooms, but are practical

Table 7-1. Dimming strategies are often driven directly by a specific need cited by the owner.

Application Considerations	Strategy	Dimming Devices
Owner needs to vary light levels during the day or after hours	Provide user or operator control of light levels	Manual dimming Preset controls and systems
Owner wants to increase worker satisfaction and productivity	Provide users with control of light levels	Manual dimming and user computer-based controls (i.e., digital/DALI)
Utility demand and kWh charges are high	Scheduled dimming and localized manual and automatic dimming	Wall-box dimmers up to centralized systems, photosensors, occupancy sensors
Energy prices can become volatile due to competitive pricing	Centralized dimming across facility to shed load, controlled by operator	Integrated systems
Utility offers incentives for voluntarily shedding load during emergency or peak demand period	Scheduled dimming across facility to shed load during peak demand periods or in response to request	Integrated systems
Daylight available from windows or skylights	Daylight harvesting	Dimmable ballasts and photosensors

anywhere manual or programmed local control can satisfy the application's needs. Wall-box dimmers are often used to provide control over light levels for spaces that require such flexibility.

The wall-box dimmer replaces the wall switch between the ballast and the power supply. First, it provides an interface that can receive commands from people (manual raise/lower dimming). Some models are preset or programmable so that control is prompted by scenes or schedules stored in memory for simple and consistent recall. The dimmer translates the commands it receives into a control signal and transmits it to a dimmable ballast, which dims the lamps. The ballasts are typically parallel-wired so that the dimmer can command all of the fixtures on the controlled circuit in unison.

Wall-box dimmers can be categorized according to depth of functionality. The basic wall-box dimmer is a manual control that enables users to control light levels. Various configurations are available including linear slides, rotary knobs and raise/lower buttons. Various types of styling are available to complement architectural interiors.

The wall-box dimmer can be combined with a manual switch for manual-on operation. These dimmers can also be deployed with photosensors, time clocks and occupancy sensor functionality to create a simple, effective, distributed lighting energy management system. This system can combine occupancy sensor switching and scheduled shut-off with daylight harvesting and manual dimming for maximum energy savings.

Integrated Dimmers

Integrated dimmers, also called preset dimmers or architectural dimmers, are programmable dimmers with a bank of preset controls (see Figure 7-1). The user can program and then recall multiple lighting scenes by pressing a button or based on a set schedule (see Figure 7-2).

Integrated dimmers aggregate a variety of features into a wall-box configuration. Commonly included features are multiple channel (zone) control, where all or selected fixtures on a circuit are controlled by the single dimmer; multiple presets; and universal circuitry that allows each dimming channel to control all types of lighting loads.

Integrated dimmers can be categorized into two levels of control. Level I integrated dimmers control a single circuit and offer multiple zone control with either a single preset or multiple presets. Level II integrated dimmers control multiple circuits and multiple zones, with either a single or multiple presets. These presets permit various lighting scenes to be created in several lighting zones around the building. Some dimmers include a built-in astronomical time clock to enable time-driven event control.

As an example, one control permits four scene preset dimming control in a single wall-box. Auxiliary controls, such as IR wireless transmitters and activators, provide remote location switching control of all or any single scene. In a commercial application, the system could create unique lighting scenes to enhance a conference room for a variety of functions, such as lectures, presentations, slide projections or meetings. Other commercial applications include private offices, restaurants, lobbies, museums and shops.

Networked Dimmers

Several manufacturers now provide communications capabilities that allow wall-box dimmers to receive commands from other devices such as keypads, master control stations, and other wall-box dimmers. These systems allow the dimming function to be distributed and thus avoid costly wiring runs back to an equipment closet. Signal communications between devices can be made with dedicated control wires between the dimmers, over the actual power conductors, or by wireless RF signals.

IR and RF Dimmers

Studies indicate that manual controls are used more frequently, and user satisfaction increases, when the user has closer access to

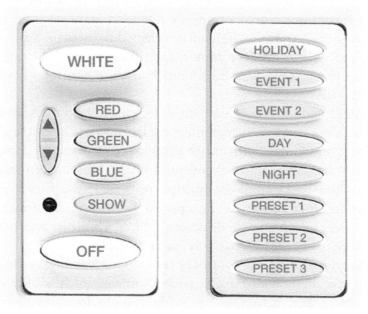

Figure 7-1. Master control station and preset wall-box dimmer. Photo courtesy of Lightolier Controls.

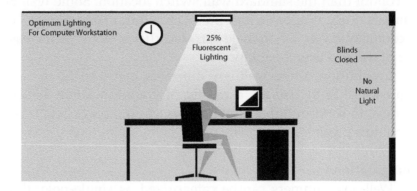

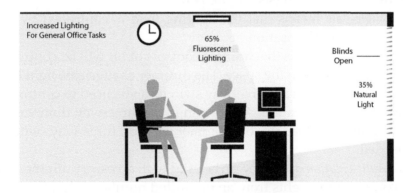

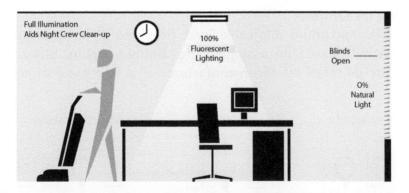

Figure 7-2a, b, c. Preset dimmers enabling the creation of lighting scenes to support typical tasks performed in the space. Graphics courtesy of Leviton.

the control than the standard wall switch location. Some wall-box dimmers, including basic and preset/programmable models, can be commanded by a handheld infrared (IR) remote for increased ease of use and user flexibility.

In addition, some wall-box dimmers may use IR or radio-frequency (RF) signals as a means of communication between dimmers to send multiple wall-box devices on, off or adjust light output to a preset scene.

Configurations

Wall-box dimmers can be categorized as single-pole, three-way or multi-location (see Figure 7-3).

Single-pole means that the lighting load is controlled by one dimmer at a single location.

Three-way means that a single control circuit will be controlled from two locations in the space. The dimmer changes the light level from one location and a three-way switch can be used to control the lights at another location. Avoid using two three-way dimmers on the same load; instead use one dimmer and add three-way switches as needed.

Multi-location dimmers can be used with accessory dimmers for full control of the lights from an unlimited number of locations.

Sizing the Dimmer

The maximum allowable load provided by manufacturers is generally based on a single-gang dimmer. Again, this rating may be de-rated by the manufacturer if a two-gang or multi-

Single-pole   **Three-way**   **Multi-location**

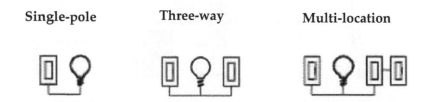

Figure 7-3. Single-pole, three-way and multi-location dimmer configurations. Graphics courtesy of Lutron Electronics.

gang configuration is used. To gang dimmers together so that the space between them is the same as for switches, a portion of the fins (heat sink) must be removed (see Figure 7-4). The removal of these fins, however, reduces the capacity the dimmer can control. The manufacturer may provide two de-rating values, based on fins broken on one side or both sides. Consult the manufacturer for more information.

It is recommended that a circuit with ballasts should not exceed 80 percent capacity. In the case of wall-box dimmers operating dimmable electronic ballasts, the manufacturer rating should be de-rated to 80 percent or less.

Power Packs

Wall-box dimmers can be operated with compatible power packs, also called "power extenders," "power boosters" or other names by manufacturers, to increase their maximum allowable load size to accommodate larger lighting loads. For some dimmers, the power pack is an option, while for others it is required to be able to operate fluorescent loads. The power pack, located on the line between the dimmer and a compatible dimmable ballast, provides on/off control and dim/bright control for the ballast. The power

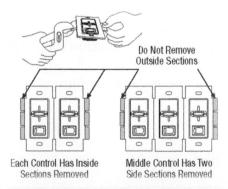

Do Not Remove
Outside Sections

Each Control Has Inside
Sections Removed

Middle Control Has Two
Side Sections Removed

Figure 7-4. To gang dimmers together so that the space between them is the same as for switches, a portion of the fins (heat sink) must be removed. The removal of these fins, however, reduces the capacity the dimmer can control. Graphic courtesy of Lutron Electronics.

pack, therefore, is essentially an extension of the dimmer but with the ability to handle a larger load. It is designed for operation with 0-10VDC (left photo) and phase-control (right photo) dimmers.

Power packs can receive commands from wall-box dimmers, occupancy sensors and photosensors while providing a power supply to these devices. It may also be programmable via the wall-box dimmer or software on a PC. The power pack is specified similarly to a wall-box dimmer in terms of rating for the load to be controlled, which is dependent on the voltage of the AC power supply. Some models feature low-end trim for setting the minimum brightness level. Otherwise, the power pack emulates the characteristics of the dimmer that it is connected to in terms of dimming range and resolution.

Compatibility

The selected wall-box dimmer (and power pack, if applicable) must be rated for compatibility with these application decisions:

- Size of the fluorescent load being controlled
- Ballast voltage (120V, 277V or other)
- Ballasts control signal range and dimming/communication method (0-10VDC, two-wire phase-control, three-wire phase-control and so on)
- Dimming range
- Single-location or multiple-location control points for the same circuit
- Single-, two- or multi-gang configuration if multiple-circuit control is required

DIMMING SYSTEMS

Dimming systems are the next step up from wall-box dimmers in terms of performance and sophistication. They offer lighting control for larger applications where wall-box dimmers and integrated systems are impractical or higher performance is required. Controls provide the user interface to match areas of control with the respective dimmers.

Typical applications include churches, restaurants, meeting rooms and multi-use facilities—anywhere higher performance dimming is required and/or where larger loads need to be controlled.

In this section, you will learn about the composition of a dimming system, the functionality of dimming system, and typical applications. You will learn about dimming panels—functionality, construction, analog vs. digital, trends in higher-end panels such as digital architecture and distributed intelligence, mounting, and sizing for the load. Finally, you will learn about control stations—functionality, features, programming, and theatrical/entertainment integration.

The Dimming System

Dimming systems consist of dimming panels (also called "dimmer racks" or "dimmer cabinets") and control stations, typically connected with low-voltage control wires. The dimming panels contain any number of dimmers capable of handling small and large loads controlling a variety of lamp types. The control stations provide programmable scene-recall, individual circuit control and scheduling.

Working together, these dimming systems allow the user to adjust the lighting to suit the task or create the proper ambience. Additionally, they can serve as the foundation for lighting energy management systems that can switch and dim loads manually and automatically.

Digital dimming systems: Another type of dimming system is a digital control system that uses DALI- or proprietary-protocol based digital ballasts (see Chapter 5). In a digital system, the ballasts are wired directly to the wall controller or other control device, simplifying wiring and eliminating the need for individual dimmer modules and dimming control panels. This section will focus on control systems that require dimmer modules and control panels.

Dimming Control Panels

A dimming panel constitutes an enclosure, a processor that assigns zones (also called channels) to lighting circuits, wiring terminations, dimmer modules (dimmers), and outputs to control the connected loads (see Figure 7-5). The panel may be analog or digital (not to be confused with DALI in this case) in construction (see Table 7-2).

If the panel is installed in the electrical room, it can be specified to include overcurrent protection (breakers).

Basic Panel Functions

Dimming panels supply power to the lighting loads and can adjust their output. Different sizes are available to manage various sizes of loads. Most dimming panels can operate almost any dimmable ballast and a variety of lamp types. The panel can be installed in the electrical room adjacent to the lighting distribution panel or as a replacement for the panel (if specified with breakers), or closer to the load (above the ceiling or in a closet). Below are the panel's basic functions.

- To provide an enclosure to house the dimmer modules.
- House high densities of dimming and non-dimming circuits.
- Provide a wiring termination point.
- Isolate the dimmers from acoustically sensitive spaces (dimmers can generate noise).
- Provide a power disconnect for each dimmed/non-dimmed circuit.

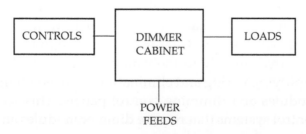

Figure 7-5. Dimming control panel. Graphic courtesy of Leviton.

Table 7-2. General comparison of analog and digital dimming panels.

Method	Analog	Digital
Feature set	"Lower end"	"Higher end"
Programmable	Yes	Yes
Intelligence	Centralized in the dimming panel microprocessor	Centralized or decentralized to control stations for greater reliability
Controllers	Slide controls	Preset/programmable integrated controls, with multiple scenes that can be programmed, stored in memory, and recalled; allows use with various protocols such as Ethernet (TCP/IP), RS-232, RS-485, Bluetooth, DMX-512, ASCII
Interfaces	Slide controls	New interfaces possible such as LCD touch-screens
Time clock integration	No	Yes, which can allow programmable scheduling
Designate one controller as master controller of others	No	Yes
Communication	Limited to connected control stations	Can communicate with other building systems via DMX-512, RS-485, RS-232, Bluetooth, Wi-Fi, contact-closures, ASCII, Ethernet, resulting in easier integration
Feedback	No	Load size/type, short circuit, overload and temperature using LED indicators on unit or connected PC software
Connectivity	Typically one pair of wires per zone	One network control wire per dimming panel or system, resulting in easier wiring
Cooling	Fans or convec-tion-cooled	Fans or convection-cooled
Maximum Allowable Load per Circuit	2400W (commercial applications, per NEC)	2400W (commercial applications, per NEC)
Size and weight	Larger, heavier	Smaller, lighter equipment allows for higher dimmer densities within the dimming panel
Noise	"Bee hive" humming	Quieter

- Enable programmable scheduling if the panel includes an internal time clock (intelligent panel).

Table 7-3. General comparison of dimming systems and stand-alone dimmers.

Method	Dimming System	Stand-alone Dimmer
Advantages	- Control of multiple control zones - Multiple adjustable settings - Wider range of calibration and adjustment	- Simpler installation - Basic operation - Generally lower cost
Dis-advantages	- Generally higher cost - More extensive installation	- Can only control one zone - Less adjustment flexibility
Ideal Applications	Larger rooms or facilities with several zones of control within each room, such as open office areas, classrooms, banquet/convention/meeting rooms and hospitality	Small spaces in a building where dimming is used in select rooms or groups, or where daylight contributions are relatively even or balanced

New Features

Some newer dimming panels are available with some or all of these features:

- Modular construction into which plug-in dimmer modules can be installed to economically size the system in direct proportion to the application need

- Designed for a higher density of dimmers due to greater compactness in dimmer modules

- Feature sections to allow 120V, 277V and emergency feeds all brought into a single panel, potentially consolidating the number of panels required and making field wiring easier by avoiding conduit runs between multiple panels

- Integrate low-voltage switching to enable programmable switching and dimming of loads

- Distributed intelligence

Distributed Intelligence/Decentralization

Some digital dimming systems provide the capabilities of distributed intelligence or decentralization. Distributed intelligence places the processing power and configuration settings into each connected control station, eliminating the need for a central processor. The panel processor simply provides an interface between the lighting controls and the dimmers. This enables the system to maintain light levels in case of a panel processor failure. If the panel processor fails, each dimmer will retain the current light level. Decentralized systems are therefore more reliable and robust.

Mounting and Connections

Dimming panels are installed in the electrical closet, mounted either on the floor or the wall (see Figure 7-6). The larger the number of lighting circuits to be controlled by the panel, the larger it is. Due to digital technology, panels are becoming smaller and lighter, allowing for higher dimmer densities to be installed within the dimming panel.

The dimming panel is wired to control stations via low-voltage wires. Some manufacturers use a dedicated power wire and a dedicated communication wire, while others integrate communication and power in a single wire. The dimming panel is also wired to the dimmable ballast via low-voltage or line-voltage wires, depending on the ballast's dimming method. Other control devices such as occupancy sensors and photosensors are typically wired to the panel to control the dimmer, which in turn controls the ballast.

Networking

Dimming and switching panels can be networked together for centralized control from a PC, which enables the facility operator to

Figure 7-6. Dimming control panels are installed in the electrical closet, mounted either on the floor or the wall. Photo courtesy of HUNT Dimming.

schedule switching and dimming of zones for load shedding and other energy management purposes.

Sizing the Panel

Dimming panels are typically sized according to number of circuits that need to be controlled, which determines the number of control outputs needed. Small dimming panels for distributed dimming (locating small panels near the loads) may contain 1, 2,

---

Dimming System Design Questions

- What do you want the system to do once it's installed? How complex is the application need?
- What is the alternative?
- How many zones will be controlled?
- What is the connected load and load type for each zone?
- How many master control and other control stations will be needed?
- Where will the master control and other control stations be located?
- What other integration must be achieved—such as theatrical consoles, A/V, shades/blinds, etc.—and how? Are these systems compatible and how?
- Are there emergency lighting circuits?
- Where are the dimming panels going to be located?
- What is the budget?

4, 4 or 12 control outputs. Typically, these dimmers have a limited capacity of 20A that is distributed between the control outputs— e.g., 1x20A, 2x10A, 4x5A. In this scenario, the size of the branch circuit conductors must be maintained for primary and secondary feeds.

Larger dimming panels for centralized dimming are sized with more outputs for control of greater number of loads—such as 16, 32, 96, etc. If the size of the load exceeds the selected panel, additional panels can be installed and wired together. For these larger applications, panels can be bused together for integrated control of hundreds of circuits. For switching control, it may be economical to wire the dimming panel to a separate switching panel if switching capability will entail an additional panel or a larger panel, or it may be more economical to mix switches with dimmers if the panel has spare capacity.

Control Stations

The control station is the "brain" of the dimming system and tells the dimming panel and its dimmers what to do.

Some control stations are available in thin-profile designs that fit in a standard wall-box, but control stations are not standard wall-box dimmers. Wall-box dimmers are stand-alone units that control

Table 7-4. Building blocks of energy-saving dimming control system.

Add ...	Receive ...
Dimmable Ballast	– Dimming capability
Wall-box dimmer (basic, preset)	– Dimming in one space – Task tuning by users – Programmable scenes – Energy savings – Mood setting – Worker satisfaction
Photosensor	– Daylight harvesting – Adaptive compensation
Occupancy sensor	– Automatic shutoff – Energy savings
Dimming panel	– Dimming one or many spaces or the entire building – Scheduled dimming (and automatic shutoff depending on model) – Load shedding
Digital, distributed dimming controls	– All of the above plus individual fixture control and networking without the need for panels
PC, Software, Networking	– Build lighting communications network connecting all controllers – Centralized control for building or buildings (with local overrides) for facility operator – Energy information

←Lower—————— Energy Savings + Flexibility + Complexity + Cost ——————Higher→

a single group of lights (with multiple units fed by the same circuit able to control several branches of the circuit), while control stations are interfaces between users and the dimmers in a system panel.

Both control stations and wall-box dimmers may offer the ability to program dimming scenes for later recall. The control station, however, may offer the ability to control multiple lighting circuits from one unit, which can be simpler and more visually pleasing than a row of dimmers on the wall. Some control stations also enable integration with screens/blinds, time clocks (for scheduled dimming and/or switching), occupancy sensors, security and building management systems, and other systems.

Features

A given control station may offer some or all of these features:

- Lighting scene preset and recall
- Individual zone intensity indication
- Individual zone intensity raise/lower buttons
- Master raise/lower buttons (causing all zones to raise or lower simultaneously)
- Scene select buttons
- Fade rate adjustment button
- Zone description labeling

Programming

Control stations come with a number of configurable buttons. Each button may be programmed for preset scene, on/off, fade, and momentary or maintained contact. Some come with an LED that lights when the unit is active.

Button configurations are factory preset to job specifications or may be set on-site by the contractor, facility operator or user. Some control stations feature an infrared receiver, which allows programming and control via an IR remote control. Other control stations come in the form of programmable LCD touchscreens.

Master Control Stations

Some digital dimming systems allow one control station to act as a master controller for the system. This means that each control station can be individually programmed and controlled by users, or all can be programmed and controlled as a group by a single

Table 7-5. Dimming controls and their application.

	Primary Equipment	*Application*
Basic	Wall-box dimmers (and optional photosensors	Dimming in a single space
Intermediate	Dimming systems (panels and control stations)	Dimming (and switching) in larger applications such as multiple rooms or buildings
Advanced	Integrated systems (networked lighting control system); digital ballasts and controls	Comprehensive lighting energy management for buildings

control station accessible to the facility operator, typically found in partitioned spaces such as ballrooms.

Theatrical Dimming Integration

Some dimming systems allow two types of control stations to operate in parallel with the panel and alternate use based on current application needs. Specifically, the panel is controlled by a standard control station during normal use of the space, then switches to control by a more sophisticated control station—such as a DMX512 theatrical lighting control console—when required. A typical application is a house of worship, ballrooms, performing arts centers and school auditoria. This type of parallel control station approach can be adapted to provide redundancy for critical dimming applications such as boardrooms and total building control.

DIMMING CONTROL ZONES

Establishing control zones is an essential step in matching equipment to the application. A control zone is a fixture or group of fixtures controlled simultaneously as a single entity by a single controller. Control zones may be called "channels" by manufacturers in their guides and equipment specifications. Zones are typically established based on types of tasks to be lighted, lighting schedules, types of lighting systems, architectural finishes/furnishings, the desired layering of the lighting, desired control flexibility and daylight availability. Zones should be established, if possible, not only on the immediate use of the space but also anticipated future uses.

Zones are often limited to the fixtures on one circuit or sub-circuit, or switch-leg, but similar fixtures on multiple circuits up to large loads can be zoned to be controlled by a single controller as well. Establishing smaller zones increases control resolution and flexibility but also increases cost. For zoning at the circuit or sub-circuit level, dimming controllers can be economical. For zoning at the ballast or fixture level, digital (DALI-based and proprietary) lighting systems can be economical, which eliminate the need for a control panel, simplify wiring, enable individual addressing of ballasts and future field reprogramming/reconfiguration without rewiring, consistent dimming performance across ballast types, and individual user control of his or her local lighting via workstation PC. The right approach will depend on the application.

In daylighting applications, it is often sound design to establish zones based on consistent task illumination needs, daylight levels and use of blinds and similar shading devices. For example, in a typical daylighting application with side-lighting, dimmable ballasts and a photosensor, each row of light fixtures parallel to the windows should be circuited and zoned separately to accommodate different daylight contributions. If the occupants along the perimeter can adjust daylight penetration via blinds, then it will be further advantageous to zone each fixture separately. For example, if one occupant closes the blinds and another opens them fully, the lighting system will provide sufficient lighting to both independently based on individual need and conditions.

Another example of small control zones is a lighting scheme with direct/indirect fixtures. A good design might utilize a direct/indirect fixture with the direct portion controllable by the occupant and the indirect portion controlled by the photosensor. The direct lighting component for each fixture would be zoned individually for adjustment to individual user need or preference. The indirect lighting component would be zoned together to dim in unison. With this strategy, the indirect portion will remain relatively consistent from fixture to fixture, thus reducing possible aesthetic issues in the space.

The size of a control zone is often limited by the current-carrying capacity of the connected control devices, which will limit how many ballasts can be controlled by the given controller. It is recommended that a circuit with ballasts should not exceed 80 percent capacity per the NEC. Therefore, a typical 20A breaker should not be loaded beyond 16A. The capacity may be further de-rated based on multiple ganging due to partial removal of the dimmer's heat sink during installation. Power packs for dimmers and dimming panels can be used to handle larger loads.

When placing equipment on plans, make sure the controls are easy to locate and to access. Don't put them in a closet that might be locked. Make sure the controlled lighting can be seen from the control panel or switch location. (Otherwise, occupants will have to yell to each other, "Is that good? Is it dim enough?") If the controls adapt to the normal behavior of people, they will be accepted. If not, they will be rejected and bypassed. Make the control scheme simple and intuitive for even the most basic operation. If controls aren't simple, they will not be used. Controls should make sense and provide flexibility to all users.

When locating a dimmer on plans, note that wall-box dimmers and control stations typically replace the wall switch, but dimmers may also be located on a desktop or be controllable by a handheld remote for greater convenience to the occupant. In addition, it is recommended to provide a manual on/off switch to allow occupants to turn off the lights, particularly if the dimmer does not allow the lamps to be fully turned off. Some wall-box dimmers integrate a dimmer control and an on/off switch in the same unit. These units are best placed where the wall switch would normally be.

Part III

Daylight Harvesting

Chapter 8

Daylight Harvesting

Daylight harvesting, also called daylighting control or automatic daylight dimming or switching, uses a ceiling-, wall- or fixture-mounted light sensor to measure the amount of illumination at the task surface in the space or at the daylight aperture, then signals a switch or dimming ballast to adjust light output from the electric lighting system to maintain the desired level of illumination. An effective daylight harvesting control system saves energy while being virtually unnoticed by occupants.

According to Heschong Mahone, energy savings from daylighting controls can range from about $0.50/sq.ft. to $0.75/sq.ft., depending on the characteristics of the application.

In this chapter, you will learn the steps in designing a daylight harvesting control system, understand common causes of success or failure in daylight harvesting installations, and view sample applications.

APPLICATIONS

With a daylight harvesting control system, electric lighting is increased or decreased in direct or approximate proportion to the amount of natural light present, resulting in a minimum maintained illumination level in the controlled space. Daylight harvesting controls can be effective in virtually any type of facility where the lights operate much of the time and where a significant quantity of daylight is provided with windows and/or skylights.

Spaces with skylights, and corridors, private offices and open cubicles near windows—particularly those with task lighting—

Figure 8-1. Successful controls in this elementary school building results in all lights off except for emergency lighting during times of day when ample daylight is available. Photo courtesy of Smiley Glotter Nyberg Architects, Inc.

are good candidates for daylight harvesting. If the entire space is uniformly skylighted, energy savings can accrue on the entire lighting load. More commonly, they apply only to the perimeter zone of a vertically glazed installation.

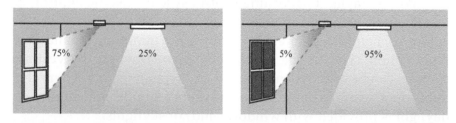

Figure 8-2. Simple side-lighting application that illustrates principle of daylight harvesting with an open-loop photosensor mounted on the ceiling. Graphic courtesy of Leviton.

ENERGY SAVINGS

Lighting system energy savings as high as greater than 90 percent have been documented in regions with ample daylight. The New Buildings Institute states daylight harvesting systems can generate maximum potential savings of 35-60 percent, but can "easily save 10-50 percent of annual lighting energy" in suitable spaces. The Lighting Design Lab states savings can reach 40-60 percent but some spaces—such as offices, classrooms and gymnasiums—can save 60-80 percent. According to DoE, daylight-response switching coupled with skylights has demonstrated energy savings in warehouses of 30-70 percent. Note that most energy savings estimates in studies indicate energy savings in the controlled lighting system only. See Table 8-1.

Since periods of maximum daylight harvesting potential correspond with periods that experience maximum air conditioning demand, daylight harvesting controls can limit peak energy demand as well save large amounts of energy. In new construction, cooling equipment can often be downsized by up to 5 percent for zones with daylight harvesting controls, according to the Washington State University Energy Program.

DAYLIGHT HARVESTING CONTROL SYSTEM

Automatic daylight harvesting control systems are comprised (see Figure 8-4 for an example):

The *electric lighting system*—lamps, ballasts, wiring to the fixtures, number of fixtures per circuit, and fixture placement and spacing.

Photosensor—ceiling-, wall- or fixture-mounted device that automatically measures light level entering the space or at the task surface, and signals the controller when a threshold is reached (light levels are increasing or decreasing).

Controller—a control unit, such as a dimmable ballast or low-voltage relay, that receives the photosensor signal as an input and

Figure 8-3. In this office space, daylight contribution is monitored, adjusting light output of the fluorescent lamps in accordance so as to generate energy savings. As daylight contribution increases, the lamps dim. As daylight contribution decreases, the lamps brighten. Photos courtesy of Leviton.

Table 8-1. Estimates of potential energy savings using daylight harvesting.

Organization	Estimated Potential Energy Savings through Effective Daylighting	Source
DoE's Federal Energy Management Program (FEMP)	Up to 75-80%	FEMP Newsletter, March/April 2002
U.S. Green Building Council (USGBC)	50-80%	USGBC Sustainable Building Technical Manual
New Buildings Institute	Maximum potential savings of 35-60%, but says automatic daylight harvesting systems can "easily save 10-50% of annual lighting energy" in suitable spaces	Advanced Lighting Guidelines
New Buildings Institute	Private office side-lighting with photosensor—up to 35% Open office side-lighting with photosensor—up to 40% Classroom side-lighting or top-lighting with photosensor—up to 40% Big box retail top-lighting with photosensor—up to 60%	Advanced Lighting Guidelines
U.S. Environmental Protection Agency	40+%	Green Lights Program
Florida Energy Conservation Assistance Program	Documented average daytime lighting energy savings of 93% in 29 Florida businesses January 2001	Energy Design & Construction Magazine,
Joel Loveland, Daylighting Director for BetterBricks Daylighting Lab	Effective daylighted buildings can be built for costs typical of common construction, and can save 40-60% energy with daylighting controls. Offices, classrooms and gyms can save 60-80%.	Lighting Design Lab News, Winter/Spring 2003
Wisconsin Daylighting Collaborative	Cool daylighting can reduce electricity consumption by lighting, fans and cooling systems by more than 50%	*Energy User News* Magazine, April 2001

issues a command to connected dimming or switching controls to adjust light output accordingly.

Dimming or switching controls—devices that receive the command signal from the controller as an input and as an output adjusts the light output of the controlled electric lighting system by dimming or switching.

Note that control components may be mounted in the

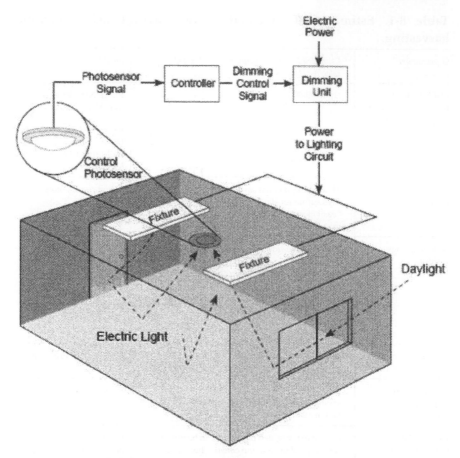

Figure 8-4. A daylight harvesting dimming system. Graphic courtesy of Lawrence Berkeley National Laboratory.

application as separate units or can be consolidated into packages; some dimmable ballasts, for example, can be matched to photosensors that directly control the ballast without the need for additional controls.

DESIGNING FOR DAYLIGHT

Daylighting is the use of daylight as a primary source of general illumination in a space. Daylighting has become a more important feature of mainstream construction due to the sustainable design movement.

Numerous studies over the last 50 years attest to the importance of daylight in design. Research indicates that daylight can improve user satisfaction/performance and retail sales. These characteristics can make daylighted buildings more valuable and marketable.

Because of the dynamic nature of the sun and volume of variables involved, daylighting design is both an art and a science, requiring diligence, expertise and commitment across the design team.

Importance of Daylight

Daylighting can impact people and spaces by providing sensory variability, connection to nature, time/weather information, full-spectrum light, modeling and an indirect component of light producing wall- and ceiling-washing effects, which can provide a more pleasant and comfortable visual environment (see Table 8-2).

Many of these benefits boil down to simple mental stimulation due to moderate changes in the environment, so long as these changes are meaningful and patterned.

Daylighting and Energy Codes

Originally, the ASHRAE 90.1 standard energy code defined daylight zones and required separate control of electric lighting in these spaces. However, these provisions were eliminated in ASHRAE/IES 90.1-1999 and are also not included in 90.1-2001 and 90.1-2004. Daylight harvesting is currently not mandated by the national energy standard.

California's Title 24-2005 energy code recognizes the importance of daylighting as an energy-saving asset instead of a heating/cooling liability. A new prescriptive provision requires skylights in "big box" buildings, specifically skylights with controls to shut off the lights when daylight is available.

The provision applies to low-rise buildings >25,000 sq.ft. under roof with >15 ft. ceilings, and with general lighting power density of >0.5W/sq.ft. For these spaces, at least one-half of the floor area

Table 8-2. Human factors research related to daylighting.

Windows and Offices: A Study of Office Worker Performance and the Indoor Environment – CEC PIER 2003 (Heschong Mahone)	Two studies looked at 100 workers in incoming call center and 200 other office workers at the Sacramento Municipal Utility District. Output was measured by a computer system and short cognitive assessment tests.	Call Center workers processed calls 6-12% faster when they had the best possible view versus those with no view. Office workers performed 10-25% better on tests of mental function and memory recall when they had the best possible view versus those with no view. Office worker self reports of better health conditions were strongly associated with better views. Workers in one study with the best views were the least likely to report negative health symptoms. Reports of increased fatigue were most strongly associated with a lack of view.
Daylight and Retail Sales – CEC PIER 2003 (Heschong Mahone)	A retailer allowed Heschong Mahone to study 73 store locations in California from 1999 to 2001. Of these, 24 stores had a significant amount of daylight illumination, provided primarily by diffusing skylights.	Consistent with a previous PG&E study, a maximum effect of 40% increase in sales was determined, putting daylighting's influence on par with parking area, competition, demographics as a predicator of retail success. Conservatively, the profit from increased sales was worth at least 19 times more than the energy savings. Employees in daylighted stores reported higher satisfaction.
Daylighting in Schools – PG&E 1999 (Heschong Mahone)	Heschong Machone acquired student performance data from three elemental school districts and looked for a correlation to daylight provided by each student's classroom environment.	In one district, Heschong Mahone found that students with the most daylighting in their classrooms progressed 20% faster on math tests and 26% on reading tests in one year than those with the least. In the two other districts, students in classrooms with the most daylighting were found to have 7-18% higher scores than those with the least.

must be daylit using skylights. The skylights must have a glazing material or diffuser that effectively diffuses the skylight. Exceptions include theatres, museums, refrigerated warehouses and Climate Zones 1 and 16.

Daylighting and LEED-NC v.2.2

Daylighting can result in multiple impacts in the LEED green building rating system but is directly addressed in Credit 8.1: Daylight and Views (1 point). This credit requires that 75 percent of all critical visual task occupied space must achieve a daylight factor of 2 percent, and occupants in 90 percent of regularly occupied spaces must have direct line of sight to vision glazing.

Architectural Design and Daylighting

Good daylighting design enables diffuse daylight to serve as a primary source of general illumination, provides gentle uniform light throughout the space, and enables occupants to control the daylight. Getting daylight into a space is not difficult—controlling it is the challenge. Good design avoids glare, direct sunlight penetration and too much daylight. An excess of direct sunlight, as opposed to diffuse daylight spread as uniformly as possible throughout the space, can cause glare and contrast problems, heat gain and result in lost energy savings opportunities.

Good design integrates daylight with the electric lighting system so that the electric lighting supplements the daylighting and operates only when needed. This is how daylighting can save energy—by using the sun as a free, abundant light source, with the building's architecture functioning as the light fixture.

Daylight can be captured through side-lighting (e.g., windows with overhangs, shading, light shelves and/or vegetation) and top-lighting (e.g., skylights, clerestories, sawtooth clerestories and monitors), which enable the architect to control the quantity and quality of light that enters the space. Generally, daylight should enter the space from as high a point as possible. Perimeter zones can be increased to maximize usable daylighting area. The Betterbricks Daylighting Lab recommends that all windows be shaded with fixed architectural elements—light shelves, overhangs, shades or vegetation.

While daylight is highly desirable in a building as a source of illumination, direct sunlight generally should not be allowed anywhere in the building except circulation areas, however. Glazing and/or shading can be used to diffuse the light as broadly and uniformly as possible throughout the space. Side-lighting can be enhanced by using sloped ceilings and lighting a space from two sides. To reduce direct glare from windows, automatic or manual blinds and louvers can be used. Windows on all sides of the building except the north side should be shaded. Look for other possible solar glare situations and control them as well.

Consider integrating automatic lighting control with automatic window shades, blinds or other devices that can reduce direct glare and heat gain (see Figure 8-5). Using the same control station, users can control both daylight and electric light levels.

Lighting Design and Daylighting

Electric lighting systems are designed to accommodate a range of tasks in the space and may perform several jobs: 1) task lighting, 2) ambient lighting and 3) accent lighting. Daylight that consistently enters the space generally provides ambient lighting and is integrated with that part of the electric lighting system.

Since daylight often does not enter the space uniformly, visualize it as gradients in a pattern. Design lighting circuits parallel to the daylight contours to enable sufficient granularity in the daylight harvesting control so that it as closely as possible imitates the daylight contribution in the space. Match the daylight

Figure 8-5. Automatic window shades used to reduce direct glare and heat gain while retaining a view. Photo courtesy of Lutron Electronics.

where people need it most.

The electric lighting system should be designed so that it places light on the same surfaces as the daylight. While studies suggest that occupants prefer light levels to change over the course of the day—in particular preferring higher light levels during the day than at night—it is not recommended to allow significant changes in surface brightness.

Therefore, if the daylight architecture distributes light on walls and ceilings, the electric lighting should also do so (e.g., indirect or direct/indirect lighting). Or if the daylight architecture distributes light downward, such as from a skylight, the electric lighting should imitate this distribution, while the designer can also consider electric indirect lighting, but note that it will likely have to operate during periods of maximum daylight contribution.

Other design considerations include:

Lamps: When warm color temperature fluorescent sources (<3500K) are used in conjunction with very cool daylight (>5000K), the lights may appear yellow. To mitigate this effect, consider lamps with a neutral-white color temperature (3500-4100+K).

Surface colors: Use light-colored surfaces in the space, with brighter surfaces kept out of the line of sight of direct sunlight.

Renovations: Adapting existing buildings to daylighting can be difficult. Single-story buildings with simple roof structures are often easiest to upgrade for daylighting, particularly spaces with high ceilings.

Glare: Ensure that windows are not a source of glare, which will result in continuous use of opaque window coverings that will keep out daylight and limit energy-saving opportunities with daylight harvesting controls.

DESIGNING A DAYLIGHT HARVESTING SYSTEM

- Step 1: Select the control method (dimming vs. switching)
- Step 2: Select the degree of automation (manual vs. automated)

- Step 3: Select the control method (open vs. closed loop)
- Step 4: Select the control method (system vs. stand-alone)
- Step 5: Select the photosensor
- Step 6: Establish control zones
- Step 7: Place photosensors
- Step 8: Place controllers
- Step 9: Establish set-points
- Step 10: Integrate the daylighting controls with other controls
- Step 11: Specify the control system
- Step 12: Commissioning
- Step 13: Occupant acceptance

Step 1: Select Control Method: Dimming Versus Switching

The first step in designing a daylight harvesting system is to select the control method. Two control methods are available, dimming and switching (see comparison in Table 8-3).

Dimming: Dimming is continuous over the dimmable ballast's range, allowing a wide range of light output.

Dimming is relatively expensive due to the inclusion of dimmable ballasts, but the smooth, gradual changes in light output are not intrusive to occupants, important in some applications.

Switching: Switching may be bi-level, with selection of three conditions—on, 50 percent light output and off—based on separately circuiting ballasts in each fixture or separately circuiting select light fixtures, or multi-level (also called stepped dimming), with selection of four conditions—on, 66 percent, 33 percent and off—based on separately circuiting ballasts operating the lamps in three-lamp fixtures.

Although the cost of dimmable ballast is falling, dimming can cost about twice as much as switching, but dimming is preferable for many applications where it is more likely to be acceptable to occupants. Sudden changes in light output can be disruptive, limiting application. If switching is selected and occupant acceptance is an issue, multi-level switching may be preferable because it offers smaller changes in light output. According to the New Buildings Institute, in high-ceiling applications, users generally do not notice changes in light level that are less than one-third of the current light level.

Table 8-3. General comparison of dimming and switching in daylight harvesting control applications.

	Dimming	*Switching*
Typical Equipment Types	Photosensor, dimmable ballast, dimmer, dimming panel	Photosensor, switch (usually low-voltage relay), ballast, control, power-packs (if individual fixture control/zoning by fixture is desired)
Light Levels	Continuous dimming from 100% to 5-10%, or stepped dimming from 100% to one or more levels	ON/OFF switching (100% and 0%), bi-level switching (100%, 50% and 0%) or multi-level switching (100%, 66%, 33% and 0%) by switching groups of lamps or fixtures (by circuit) or individual fixtures (using a power-pack)
Advantages	Smooth transitions between light levels, greater control accuracy, more options for integration with personal control strategy, greater occupant acceptance, higher energy savings in spaces where daylight levels are highly variable and/or close to electric light levels	Lower initial cost (although more ballasts may be required than equivalent dimming strategy), easier to commission, higher energy savings in spaces with high, consistent amounts of daylight
Dis-advantages	Higher initial cost, dimming ballasts less efficient than the most efficient non-dimming ballast, more sophisticated commissioning	Less occupant acceptance, less control accuracy
Occupants	More effective in spaces with occupants performing stationary or critical tasks such as offices, where changes in light output should be virtually unnoticed	More effective in spaces with occupants performing non-stationary, non-critical tasks such as hallways and atria, where abrupt changes in light output will be minimally intrusive
Daylight Levels	More effective if daylight provides only portion of required light level in space or daylight contribution is variable; dimming can respond to fluctuating conditions without being intrusive	More effective if daylight provides consistent, large contribution (2-3 times greater than minimum electric light levels); in climates with clear skies, fixtures can be shut off and remain off for hours
Lamp and Fixture Visibility	Fixtures can be mounted in normal field of view, lamps can be visible to occupants	More effective in spaces where fixtures are mounted high and not in normal field of view; occupants should have limited view of lamps and instead primarily view illuminated surfaces, such as with indirect fixtures

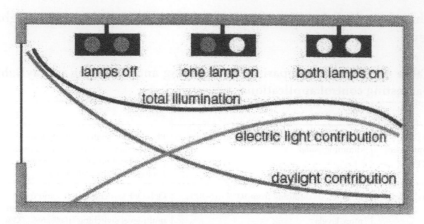

Figure 8-6. Switching in a side-lighting daylight harvesting application. Graphic courtesy of Lawrence Berkeley National Laboratory.

Step 2: Select Degree of Automation

The next thing to decide is who or what is going to tell the controller to dim or switch the lights. Two degrees of automation are available, automated and manual.

Manual control requires user intervention to switch or dim the lights based on preferred light levels, with daylight contribution a factor in the user's decision-making. Automatic control utilizes a photosensor to monitor light levels and automatically adjust light output. With manual control, no photosensor is required and occupants are likely to accept the controls, but energy savings are typically significantly lower than the potential of automated systems because human initiative is required. There are also limited options for occupant control—generally, the bi-level or multi-level switching will be activated via a wall switch.

With automatic control, a photosensor is required and the control method must be appropriate for the application, controls properly calibrated, and users educated about the controls to ensure occupant acceptance. However, energy savings potential can be higher. When dimming is used, there are also more options for individual user control of the lights, which can enhance savings while promoting user acceptance.

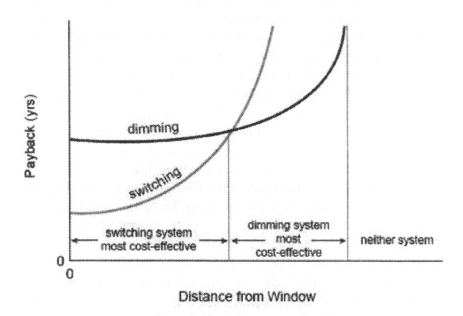

Dimming/Switching payback chart

Figure 8-7. Switching versus dimming in a side-lighting daylight harvesting application. Graphic courtesy of Lawrence Berkeley National Laboratory.

The remainder of this chapter focuses on automatic daylight harvesting control systems.

Step 3: Select Control Method: Open Versus Closed Loop
Daylight harvesting controls may be "open loop" or "closed loop" systems. They measure the daylight contribution on the task surface differently.

Closed-loop systems measure the combined contribution to light level from both daylight and the electric lighting system, then adjust light output to maintain the desired level of illumination. Because the photosensor measures the electric lighting system's light output, it "sees" the results of its adjustment and may make further adjustments based on this feedback—creating a closed loop. See Figure 8-8.

The primary advantage of closed-loop systems is that they pose a lower initial cost when only a single zone must be controlled, and, unlike open-loop systems, they measure actual light level by sampling the task surface, so they will respond to users opening and closing blinds and other changing conditions.

Closed-loop systems are generally economical for control of smaller spaces, or larger spaces with all the lights controlled in a single zone. An example is a private, windowed office. Generally, they provide a lower-cost solution when a single zone is controlled. They are not recommended for control of multiple zones.

Open-loop systems measure only the incoming daylight, not the contribution from the electric lighting. The photosensor should not see any electric light and therefore it is mounted outside the building or inside near a daylight aperture. Because there is no feedback, it is an open loop. In the case of a switching system, the photosensor signals the lights to shut off when daylight reaches a predetermined level. In the case of a dimming system, the photosensor measures incoming daylight and signals a controller to proportionately dim the lights based on the estimated daylight contribution. See Figure 8-9.

The primary advantage of open-loop systems is that they are able to control multiple zones from a single photosensor, as opposed to closed-loop systems, which require that each zone

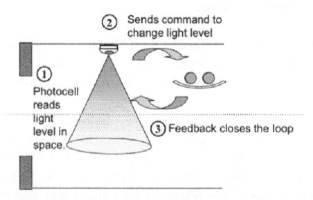

Figure 8-8. Closed-loop photosensor. Graphic courtesy of Watt Stopper/ Legrand.

be controlled by a dedicated photosensor. (In review, a zone is a fixture or group of fixtures that are controlled simultaneously.) In addition, open-loop systems provide greater calibration flexibility than most closed-loop systems, and are more "forgiving" to errors in placement of the sensor or its field of view.

Its primary disadvantage is that the system responds only to exterior daylight availability and not actual daylight contribution in a space; if an occupant closes the blinds, the system will not recognize that and dim the lights anyway. For this reason, local overrides are useful.

In dimming applications, open-loop systems are generally economical for control of larger areas with multiple adjacent control zones, because a single photosensor can be used for control of multiple control zones. An example is an open office. In applications with a single zone, open-loop systems generally pose a higher initial cost than closed-loop systems. Open-loop systems are also recommended for high-bay applications with skylights, as the photosensor can be mounted in the lightwell of the skylight, while with a closed-loop system, it may be difficult to find a good photosensor viewing location.

Step 4: Select Control Method: System Versus Stand-Alone

Daylight harvesting controls can be configured as either systems or stand-alone controls. Control systems include controllers and

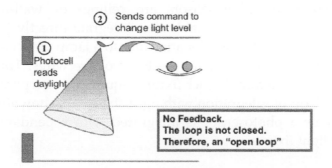

Figure 8-9. Open-loop photosensor. Graphic courtesy of Watt Stopper/ Legrand.

external photosensor as well as more adjustment and configuration options. Stand-alone controls are self-contained, independent units.

Control systems enable control of multiple lighting zones and offer multiple adjustable settings, ideally suited to large spaces with multiple zones, such as open office areas, classrooms, warehouses, cafeterias, etc. However, they are generally more expensive and require more extensive installation.

Stand-alone controls enable simpler operation with one low-voltage device plus a power-pack and generally a lower installed cost, ideally suited for smaller spaces where daylight contributions are relatively even and balanced, such as enclosed offices, hallways, etc. However, stand-alone controls are capable of controlling only a single zone, and offer fewer adjustment options, which can limit application flexibility.

Step 5: Select Photosensor

Automatic daylight harvesting control systems use a photosensor to measure light level on the task surface or entering the space—measuring reflected light but not direct sunlight. The photosensor is a small device that can include a light-sensitive photocell, input optics and an electronic circuit used to convert the photocell signal into an output control signal, all within a housing and with mounting hardware. The visible size of a photosensor on the ceiling ranges from a golf ball to a standard wall switch.

While many photosensors are ceiling- or wall-mounted, photosensors are now available that integrate directly into open and louvered light fixtures by attaching to a lamp via a clip and to the fixture's dimmable ballast via low-voltage wires. This allows individual fixture control; each fixture requires its own photosensor. Another form of integration with light fixtures is shown in Figure 8-10, where a photosensor is integrated with a pendant direct/indirect light fixture.

Control Algorithm

Most photosensors use one or a combination of up to three

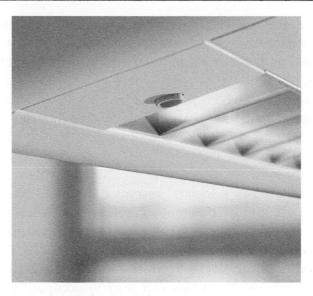

Figure 8-10. Fixture with integral photosensor. Photo courtesy of Ledalite Architectural Products.

control algorithms—constant set-point (threshold), open-loop proportional (continuous), and closed-loop proportional (sliding set-point).

Constant set-point is intended for use in closed-loop systems. The set-point, determined by the designer, indicates the threshold at which the lights will be raised or lowered. The threshold in turn is based on a target light level. The set-point control method aims to keep the photosensor signal constant during the operation of the system, somewhat like a thermostat. It has been found through practice that the proportional control method is superior to the set-point method in interior applications.

Spatial Response

The photosensor's spatial response describes its sensitivity to light from different directions and defines its field of view—what it "sees," in effect. This is determined by the design of the optical system that gathers and delivers light to it. If the field of view is too broad, the photosensor may detect light where it shouldn't, such as detecting direct sunlight near or outside a window, and thereby possibly dim the lights below what is intended. If the field of view is too narrow, the photosensor may become too sensitive to changes in brightness within a localized area, and raise or lower the lights incorrectly.

According to the New Buildings Institute, a 60-degree cone of vision is common. Some sensors provide an adjustable feature to shield direct sunlight from the field of view.

Photopic Correction

The photosensor's spectral response describes its sensitivity to optical radiation of different wavelengths. Photosensors can respond to a greater range of the electromagnetic spectrum than the visible light portion, which the human eye can see. For example, it can respond to ultraviolet and infrared radiation and thereby dim the lights unnecessarily, which can lead to occupant complaints. For this reason, filters are used to as closely as possible match the photosensor's "eye" to the human eye.

When daylight and an electric light source are mixed in the same space at different levels, the photosensor's control algorithm should automatically undergo photopic correction. Open-loop and closed-loop proportional control algorithms can accomplish this. However, when daylight and two spectrally different types of electric lighting—such as fluorescent and incandescent—are mingled in the same space, correction is not currently possible with today's technology.

Step 6: Zone for Daylight Harvesting

In review, a control zone is a light fixture or group of fixtures that are controlled simultaneously by a controller. In daylight harvesting applications, zones are established based on a combination of factors.

Perhaps most important is ensuring that the light fixtures in each zone receive a consistent amount of daylight at any given time while having consistent light level requirements. The challenge is to minimize the number of control zones, as each control zone adds to cost, while ensuring maximum response of the control system to daylight availability so as to save the most amount of energy while providing appropriate lighting.

Zoning can be accomplished in a number of ways, but one effective way to facilitate logical control zones in new construction is to plan circuit wiring for light fixtures around daylight sources such as windows and skylights. Daylight control zones should be switched separately from other zones.

When planning control zones, be sure to provide local overrides to enable users to override the daylighting control and/or automatic shut-off.

Side-lighting and Top-lighting

Generally, two types of daylight harvesting needs exist, distinguished by the distribution of daylight in the controlled area.

Perimeter zone applications are the most common since daylight enters a space through vertical windows. The distribution of daylight tends to be highly non-uniform, with large amounts in areas close to windows and rapidly decreasing amounts further away. In these situations, it is desirable to control light fixtures adjacent to the glazing separately from those further in to obtain maximum energy savings while still providing necessary task illumination. Depending on the dimming system chosen, it may be best to specify that power wiring for the fixtures run parallel to the windows rather than radially outward from the building core. This can be an important consideration in retrofit or renovation installations.

The second type of daylighting situation generally occurs in skylighted areas where the distribution of daylight is relatively uniform throughout the controlled space.

As a starting point for creating zones, try to control toplighted areas separately from sidelighted areas.

Consistent Daylight Illumination

Electric lighting systems are designed to accommodate a range of tasks in the space and may perform several jobs: 1) task lighting, 2) ambient lighting and 3) accent lighting. Daylight that consistently enters the space generally provides ambient lighting and is integrated with that part of the electric lighting system.

It would not make good sense, however, to have a controller dim the lights across an entire open office simply because the perimeter is getting a lot of daylight through the windows. In this case, we would want the flexibility to control the lights near the perimeter separately from the lights in the interior that is receiving less or no daylight.

For this reason, the ambient lighting system should be layered around the contours of daylight availability. Since daylight often does not enter the space uniformly, visualize it as gradients in a pattern.

Consistent Light Levels

Besides daylight availability, zoning is based on lighting need. Identify areas that are used for similar types of activities with similar lighting requirements.

Architectural Finishes

Besides blinds, architectural finishes such as walls, ceilings, floors, furnishings, etc. can impact lighting in the space.

Some areas may have darker finishes and others lighter finishes.

Designers often set control zones to accommodate these impacts on the space if there are significant differences from one area of the space to the next.

Sizing the Zone

It is a rule in lighting control that the higher the number of control zones in a space, the more flexibility and control accuracy will be gained, at the expense of cost and more sophisticated commissioning. In an office side-lighting application, for example, if each user on the perimeter area can control daylight penetration using window blinds, then it may make sense to separately zone each fixture in the first two rows parallel to the window. That way, if one user has the blinds up and another has the blinds down, creating different availability of daylight, their local lighting will be maintained at the desired light level. Think in terms of small zones when side-lighting.

In top-lighting applications, however, zones may be larger given areas with common skylight configurations, roof height and area function.

Aesthetics

Designing for small zones in a large, open space can result in aesthetic issues—namely, a checkerboard pattern on the ceiling. However, this can be addressed by using direct/indirect pendant-mounted fixtures in which the direct component is zoned by fixture with occupant control, and the indirect component of all of the fixtures is zoned together to dim in unison according to a signal from a photosensor.

With this strategy, the indirect portion will remain relatively consistent from fixture to fixture, thus reducing possible aesthetic issues in the space.

Zoning in a Side-lighted Space

Spaces with side-lighting can pose special challenges. The designer should be aware of factors limiting daylight penetration (the daylighted area is generally defined as about 1.5 times the head height of the window in from the window) and the appearance of shadows, particularly when using switching.

Be conservative with defining the daylighted area and accommodate shadows from other buildings, vegetation, etc. by making control zones smaller.

Confine control zones to a single building exposure and so that each includes the same window type.

Be conservative when controlling lights that are isolated from the windows by a partition that is 5 ft. or higher.

Finally, circulation spaces running along window-walls should constitute a separate control zone.

Zoning with Digital Controls

Daylight harvesting control applications may demand a high degree of granularity of control zones for optimal energy savings and performance. Zoning at the fixture level can present challenges with conventional wiring and analog control in terms of cost, commissioning and maintenance.

Emerging technology, specifically digital dimming control systems based on the DALI protocol or proprietary protocol, enable practical individual fixture control.

With digital dimming control, wiring is simplified because all fixtures are connected by a single two-wire bus, forming a lighting network of addressable ballasts that can be individual programmed or programmed in groups. The facility operator, in other words, can set the desired light level for each fixture.

The lighting contours in side-lighting and top-lighting applications can be matched to the light levels of each fixture to provide the desired visual environment. Separate scenes can be programmed for different daylight conditions—dark, overcast and clear—allowing the daylight harvesting control system to respond to the range of daylighting conditions. The ability to fade gradually from one scene to the next allows lighting changes to be virtually unnoticed.

Step 7: Place Photosensors

For a precise correlation between photosensor input and light level on the task surface, the ideal location for the photosensor is at the task surface. However, this is not practical because activities in the space would interfere with the photosensor's performance. For proper performance, photosensors are typically mounted on walls, on the ceiling or integrated with light fixtures. In side-lighting applications, the photosensor is typically placed on the ceiling looking down on a representative task area. For skylighting applications, the photosensor is often placed in a skylight looking up at the available daylight.

A leading cause of daylight harvesting project failures is improper placement of photosensors. Since the performance of the control system depends entirely on what the photosensor "sees," proper placement is critical, especially in side-lighting applications, which rely on reflected, diffuse daylight, and where direct sunlight can affect the performance of photosensors. Even a slight change in location and orientation can affect the performance of photosensors.

The challenge is to orient the sensor in such a way that is measures reflected daylight in proportion to how it varies on the task surface. The ideal placement is such that the sensor has a high level of illumination from daylight, but is shielded from any exterior glare sources.

The photosensor should be placed so that it receives a representative sampling of daylight. Too broad a field of view for the sensor can result in detecting direct sunlight or light sources outside the control zone. Too narrow a view can make the sensor too sensitive to local changes in brightness.

Use a light meter to measure light levels at potential locations before choosing the final placement.

Otherwise, placement is highly dependent on the application characteristics.

Placing Open and Closed Loop Photosensors

Closed-loop photosensors are generally mounted on the ceiling so that it views a representative area—including the lighted area that it is controlling (see Figure 8-11). They should not directly view the window or a pendant fixture. Closed-loop systems must be set up with light level readings under both daytime and nighttime or dusk (or approximating dusk—i.e., with blinds closed) conditions.

Open-loop photosensors are also typically mounted on the ceiling, but look toward the window or up into a skylight well to view incoming daylight but not the lighted area being controlled (see Figure 8-12). Open-loop systems tend to be easier to set up, requiring a light level reading only during the daytime.

Regardless of technology or product selected, follow the manufacturer's instructions for placing a photosensor.

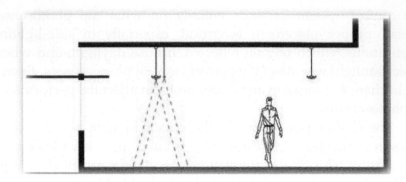

Figure 8-11. Placement of closed-loop photosensor. Graphic courtesy of Lighting Design Lab.

Figure 8-12. Placement of open-loop photosensor. Graphic courtesy of Watt Stopper/Legrand.

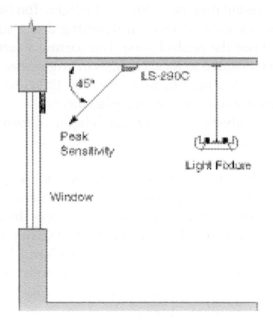

Ten Tips for Placing Photosensors

Photosensors should be located carefully to synchronize the availability of daylight with coverage from the electric light fixtures.

Avoid placing a photosensor on a roof for indoor lighting control or indoors at a wall switch location.

The photosensor should be placed so that it receives a representative sampling of daylight.

The location of the photosensor should be indicated on the bid documents.

If the space has only one primary task area, mount the photosensor above the task.

In side-lighting applications, a rule of thumb is to place the photosensor about two-thirds of the depth of the control zone away from the window.

Note that a location selected based on the recommendation of one manufacturer for its own product may not work on another manufacturer's product. Carefully read and follow the photosensor placement guidelines in your design for the selected product and manufacturer.

If a skylight is used, consider placing the photosensor under the skylight glazing rather than above it.

If a light shelf is used, consider mounting the photosensor above the shelf so that it directly views daylight reflected from the shelf.

Photosensors should not directly view pendant fixtures that it controls.

Step 8: Place Controllers

Install a controller for a system with a remote photosensor as close to its control zone as possible, so as to ensure easier set-up, adjustment and ongoing maintenance.

Whenever possible, mount the controller so that the controlled lamps are visible by operators who are adjusting the controller.

Step 9: Establish Set-points

Daylight harvesting controls dim or switch to maintain a target set-point. Determining the set-point is a critical design decision that can be fine-tuned during commissioning.

The set-point should be higher than the designed maintained light level. Higher set-points can decrease energy savings but are likely to increase user acceptance. Studies indicate that people like light levels to increase as daylight contribution increases, that their awareness of switching or dimming decreases as overall light level increases, and that they tolerate light level dynamic lighting changes typical with daylighting as long as their task lighting needs are met. Typical daylighting practice allows daylight levels to be 2-3 times higher than the design light level, so higher levels of daylight are likely available.

In dimming applications, recommended practice is that dimming not begin until the daylight contribution is 150 percent of the design light level. In switching applications, recommended practice is that switching not occur until the daylight contribution is 150-200 percent of the level to be reduced.

Set-points and Switching

Switching interior lighting as a function of available daylight is inexpensive but can be intrusive. For this reason, photoswitches should provide switching at "safe levels" with a wide deadband to minimize nuisance switching. Some photoswitches are also available with time delays to further minimize nuisance switching by ensuring that the system is slow to respond to sudden daylight changes. These operating parameters are set in the field during calibration of the daylight harvesting control system. In this sense, photoswitches are like occupancy sensors, with adjustable sensitivity and time delay response, that respond to measured light levels instead of occupancy. See Figure 8-13.

Photoswitches should have two set-points, which are separated by deadband, and two time delays. The deadband defines sensitivity of the switching response to daylight by creating a buffer between the threshold levels (ON set-point and OFF set-point). The shorter the deadband between the two set-points, the greater the sensitivity. If the deadband is too small and daylight conditions are variable (e.g., intermittent cloud conditions), the controls may switch on and off frequently during the day, which can reduce lamp life while annoying occupants.

When daylight increases, the monitored light level increases, first above the on set-point, but still within the deadband, so the lights stay on. As the daylight increases further and the light level rises above the off set-point, the lights are switched after the time delay and remain off until the light level falls below the on set-point and the second time delay lapses. If the photoswitch senses the electric light it is controlling, then the deadband should be larger.

Usually, the time delay for switching the lights on is very short to ensure users have adequate light levels, while the time delay for switching the lights off is longer to avoid frequent nuisance switching due to variable daylight conditions. For example, the on time delay might be 20 seconds, while the off time delay might be 20 minutes.

Determining the deadband and time delay not only requires understanding of local daylight conditions, but also a decision about tradeoffs between energy savings and occupant acceptance. As deadband and time delay increase, energy savings and the likelihood of nuisance switching will decrease.

Set-points and Closed-loop Systems

If a closed-loop system is used, the deadband must be adjusted to account for the removal or a large portion of light level that was being contributed by the electric lighting system. Otherwise, as daylight increases past the on set-point, the lights may switch, which significantly decreases light level, which—if it falls below the off set-point, will cause the lights to switch back on again. To

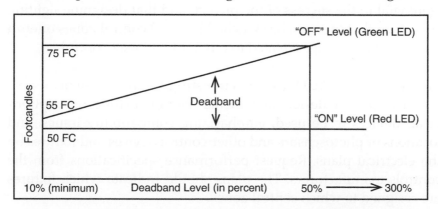

Figure 8-13. Setpoints and deadband. Graphic courtesy of Gentec.

avoid cycling, the deadband must be larger than the contribution from the electric lights that will be removed during switching.

Closed-loop systems must work with a sliding set-point control; as daylight increases, the set-point also increases, so that the light level is maintained or even increased as daylight increases.

Step 10: Integrate Daylighting and Other Controls

Daylight harvesting controls can be integrated with automatic shut-off controls such as occupancy sensors and scheduling systems for energy code compliance and/or to increase energy savings. Daylighting dimming can also be integrated with individual occupant dimming to increase energy savings and user satisfaction. However, be sure to include consideration of conveniently located local overrides over scheduled shut-off in your planning of control zones.

A case in point is the U.S. General Services Administration (GSA), which approved installation of daylighting controls, occupancy sensors, manual dimmers and scheduling devices at the Phillip Burton Federal Building and monitored the results.

In private offices, occupancy sensors and automatic daylight dimming reduced lighting energy consumption by an average of 45 percent on weekdays. In open daylighted offices, savings from the daylight harvesting control system alone were significant.

GSA discovered that proper installation and commissioning were vital to the success of the project, and that designing lighting circuits so that they can be switched in small control zones offers a number of advantages related to daylight and user control.

Step 11: Specify the Daylight Harvesting Control System

The project design documents should include descriptions of all devices to be used, emphasizing compatibility issues, and locations of photosensors and other control devices and fixtures on the electrical plans. Request performance specifications from the controls manufacturers. The plans should indicate which fixtures are assigned to which control zones.

The project design documents should also include a wiring

diagram for the controls that shows all wiring connecting them, identifying whether the wires are low or line voltage. All lighting controls, compatible ballasts and ballast configuration within fixtures should be identified. Control schematics are especially important where different buildings systems come together; for overlaps, identify responsibilities for adjustment, etc.

Finally, and perhaps most important, the documents should provide a sequence of operations for the control system, including the design intent, design light levels and control set-points. In addition to the sequence of operations, start-up, calibration and testing procedures should be provided to the electrical installer. The contractor will use these documents to commission the system. The designer can require the installer to complete a start-up form for each daylight harvesting control that is installed.

Step 12: Commissioning

Most daylight harvesting systems fail because they are not commissioned. Daylight coverage in each space is virtually unique and therefore dimming solutions typically will not meet the design intent right out of the box.

Provide specific guidelines and expectations for checkout and verification of the controls.

Specify commissioning services as a separate item, to be bid separately.

Some manufacturers provide commissioning support.

Commissioning and calibration activities related to daylight harvesting control systems can be found in Chapter 10. More detailed commissioning tips for daylighting applications with windows can be found at the Lawrence Berkeley Laboratory website at http://windows.lbl.gov/daylighting/designguide/copyright_tips.html.

Step 13: Occupant Acceptance

Discuss the controls, how occupants should use the controls, and occupant satisfaction with the owner of the building. Educate the owner and occupants about the owners during the commissioning and occupancy phases to ensure acceptance.

This is an ongoing process that may require further adjustments or recalibration of the daylight harvesting control system down the road as occupants, managers, architectural finishes, furniture or use of the space change.

At the end of the commissioning phase, give all documentation about the control system—such as specifications, warranties, construction documents and operating manuals—to the facilities staff, and provision training for the staff on operation and maintenance of the system by the manufacturer.

Application: Private Office

The application in Figure 8-14 is an enclosed single-occupant office in which primary tasks include computer work, meetings and reading. Daylight enters through a window, generating energy-savings opportunities with daylight harvesting controls in addition to automatic shut-off. In this space, automatic daylight dimming is provided using dimmable ballasts connected to a photosensor. For automatic shut-off, a ceiling-mounted occupancy sensor is also installed.

Application: Gymnasium

Imagine a gymnasium with skylights and T5HO hi-bay light fixtures, in which primary tasks include classes and games during

Figure 8-14. Private office application example. Graphic courtesy of Watt Stopper/Legrand.

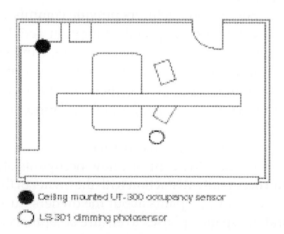

● Ceiling mounted UT-300 occupancy sensor

○ LS-301 dimming photosensor

both daytime and nighttime hours. Daylight entering through the skylights provide energy-saving opportunities with daylight harvesting control. In this space, bi-level or multi-level switching can be employed, coupled with automatic shut-off to comply with energy codes.

If the fixtures are 4-lamp, each fixture will have two 2-lamp ballasts controlled on separate circuits, enabling bi-level switching (on, 50 percent and off). If the fixtures are 6-lamp, each fixture will have three 2-lamp ballasts controlled on separate circuits, enabling multi-level switching (on, 66 percent, 33 percent and off).

Application: Classroom

The classroom shown in Figure 8-15—with primary tasks including reading, computer work, testing and presentations— receives daylight from multiple sources, presenting energy-saving opportunities with daylight harvesting control in addition to automatic shut-off. Manual override capability will also be provided for user control of light levels. The direct/indirect pendant fixtures

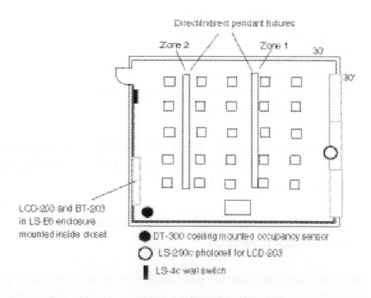

Figure 8-15. Classroom example. Graphic courtesy of Watt Stopper/ Legrand.

are divided into two zones in which dimmable ballasts enable continuous dimming. Both are controlled by a ceiling-mounted occupancy sensor and local override easily accessible to the teacher.

COMMON FAILURE MODES

David Eijadi, FAIA and Principal at The Weidt Group (www. twgi.com), conducted a study of daylight harvesting projects to find out why they succeed or fail. The Weidt Group provides energy design assistance, including daylighting as an energy conservation strategy, to architects and engineers, consulting on more than 150 projects annually. "Design analysis often shows daylighting control to be one of the most promising energy conservation strategies for commercial buildings; consequently, daylighting controls are more frequently installed," says Eijadi.

The study concluded:

- Savings from automatic daylighting control systems are often not realized fully when a building is turned over to users.

- Daylighting performance needs attention and evaluation from multiple design disciplines during the design development and construction process.

- Users are not educated about the installed control systems; when something doesn't work, users often disable the system instead of getting it fixed.

The Weidt Group study then cited common examples of why daylight harvesting project fail:

- Lack of coordination or understanding between the design disciplines concerning the daylighting control system

- Improper location of daylighting controls

- Inadequate specification of the controls systems, component parameters and sequence of operations

- Shop drawings made by contractors that detail the system are not checked, or the lighting designer does not know what to check

- Field changes to tune a system are not documented and taken back to the designer to complete the feedback loop

These problems result in common failure modes, described in detail in Table 8-4:

- Under-dimming, which results in less than expected energy savings

- Over-dimming, which results in user irritation

- Frequent cycling of dimming or switching, which results in user irritation

- Lights left on at night, which results in less than expected energy savings

Table 8-4. Common failure modes with daylight harvesting control systems. Source: The Weidt Group, 2006.

Case Study	1	2	3	4
Space type	College Dining Hall	College Classrooms	Office Building	Office Building
Failure mode	**Under-dimming**	**Under-dimming**	**Over-dimming**	**Under-dimming**
Effects of Failure	Reduced energy savings	Reduced energy savings	Reduced energy savings	Reduced energy savings
Root Cause	Not wired correctly	System not calibrated	Calibrated aggressively	System not calibrated
Additional Causes	System not calibrated	Windows smaller than expected	Occupants have history of higher lighting levels	Too many sensors installed, calibration not feasible
		Sensor sees indirect lights	Dark furnishings create dark space	
Action to correct situation	Re-wire sensors to control dimming light sources	Proper calibration	Continue to test the ability of the occupants to accept some lighting control	Remove sensors from the daylighting system, control lights with 1 sensor per orientation
	Calibrate system	Educate operator		Proper calibration of remaining sensor
	Educate operator	Educate user		Educate operator and user

Case Study	5	6	7	8
Space type	College Classrooms	Big Box Retail	Office Building	Recreation Center - Pool
Failure mode	**Cycles**	**Over-dimming**	**Lights on at night**	**Under-dimming**
Effects of Failure	User irritation	Concern for store revenue to be reduced	Reduced energy savings	Reduced energy savings
Root Cause	Faulty controller	Calibrated aggressively	Night-time over-ride not available	Sensor location does not detect enough light
Additional Causes	Photosensor and controller incompatible	3 daylight zones makes calibration a more complex task to do accurately.	Wrong sensor type installed	System not calibrated
		Daylight is not uniform in the space		
Action to correct situation	Change programming in EMS system to set delay for dimming by the photosensor	New owner needs to understand system and become convinced to try re-calibration	Change sensor type and relocate sensor	Remove poorly located and not working sensors from the system.
				Control fewer fixtures with the working sensor

Figure 8-16. Vermont-based manufacturer NRG Systems in 2005 built a new headquarters carefully crafted to reflect the company's commitment to the environment, the community and its employees. The 46,550-square-foot facility, which includes office, manufacturing and warehouse space for company-produced wind monitoring equipment, was designed to minimize environmental impacts and maximize energy conservation. It serves as an example of daylight harvesting done right.

Project goals: The first design goal was to find ways to minimize energy needs in all areas of building operation, including heating, cooling and lighting. The second goal was to provide as much energy from renewable sources as possible. Company owners were willing to make an upfront investment to ensure significant long-term savings.

Naomi Miller of Naomi Miller Lighting Design specified the electric lighting and Andy Shapiro of Energy Balance, Inc. developed the facility's daylighting plan. The two designers collaborated with Watt Stopper/Legrand to select energy saving controls that would meet the ambitious criteria of the project.

Daylighting contribution and controls: Careful building orientation and design allows daylighting to provide significant contributions not only in office areas, but warehouse spaces as well. Light-guiding blinds, coupled with strategically placed windows and skylights, bring daylight deep onto the building and provide an abundance of diffuse, low-glare lighting.

Augmenting the daylight is an electric lighting system consisting primarily of high-performance, T8 fluorescent lamps with dimming ballasts controlled by Watt Stopper daylighting controls. Photosensors constantly monitor outdoor lighting levels and controllers automatically adjust the fluorescent lighting to maintain the desired ambient levels. A multi-zone control strategy is used in the office areas so that fixtures closest to the windows dim first as daylight contribution increases.

Occupancy sensors: Additional energy savings are achieved by using occupancy sensors throughout the building and grounds to ensure that lights are not on unless the space is in use. Most are set with the manual-ON option selected in order to maximize energy savings.

Passive infrared, ultrasonic and dual technology sensors have been installed according to the size and function of each space. Dual technology sensors, advantageous for applications not ideally suited to either PIR or ultrasonic technology alone, are used in several offices and a large board room. This approach ensures the greatest sensitivity and coverage with the least threat of false triggers. Both technologies must detect occupancy before the sensor turns the lighting on, but continued sensing by just one technology will hold the lighting on.

Special weatherproof sensors are used for outdoor areas and are rated for temperatures from -40° to 130° Fahrenheit—especially suited to Vermont, which is prone to very cold temperatures in the winter.

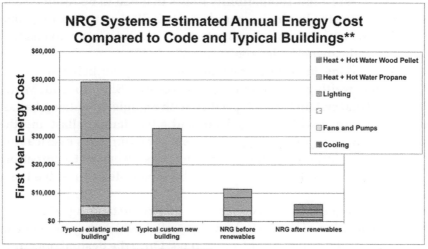

NRG Systems Estimated Annual Energy Cost Compared to Code and Typical Buildings**

* Based on energy usage of similar sized existing facility, rough estimate
** Based on computer modeling of building performance and estimates of renewables contribution

The spaces using natural lighting employ sensors along with the daylighting controls to maximize energy savings while at the same time helping NRG Systems keep its commitment to being a good neighbor. Community residents also benefit from the sensors as they reduce light pollution from the skylights after dark.

Lamp and ballast selection: NRG Systems' electric lighting primarily uses high-performance T8 lamps combined with programmed start electronic ballasts for maximum light output and long life. Low, normal, and high-output ballasts were specified for different spaces to customize the light output appropriately and squeeze every possible watt out of the lighting operation.

While this approach sounds simple, it complicated the design and construction processes. Multiple fixture tags were required for each basic fixture type, baffling suppliers who didn't understand the ballast differences, and submittal reviews became complicated. Additionally, the contractor had to be careful about putting identical-looking fixtures in specific rooms. Downlights, wall sconces, and decorative pendants primarily use 32W compact fluorescent lamps. Limiting the number of lamp types on the project simplifies maintenance and allows owner to economize by purchasing replacement lamps in volume.

System tuning: When asked whether the lighting and controls systems performed perfectly right away, the designers responded, "No. There were a few compatibility issues that had to be resolved and required getting the luminaire manufacturer, ballast manufacturer, lamp manufacturer and the controls manufacturer involved in a conversation with us and the owner's electronics guru."

After taking field measurements, observing the behavior of the system and waiting for laboratory test results, Shapiro and Miller discovered that certain fixtures specified with a single three-lamp ballast had been shipped with a single-lamp and a two-lamp ballast and the added current draw of having two ballasts per fixture was overloading the system. Additionally, the lamps provided were not compatible with the ballasts, so the fixtures had to be reballasted to resolve these problems.

The designers also discovered a misunderstanding about the standard operation of the daylighting controller, which they intended to have switch lights off on bright sunny days. Once the desired operation was communicated to engineers at Watt Stopper/Legrand, they were able to provide instructions on how to re-wire the controller to produce the desired results.

Shapiro and Miller recalled that the electricians experienced some difficulty implementing systems where fixtures were controlled based on input from both daylight and multiple occupancy sensors due to the complexity of the resulting sensor/relay/ballast/switch wiring diagrams. They noted, "Commissioning the sensors took a significant amount of time, particularly given the owners' desire to minimize electricity usage, which required repeated adjustment of sensitivity and off-delay time settings. However, the time in commissioning paid off. The systems are now working as expected."

Design achievements: All of this careful attention to detail resulted in a lighting design that was estimated to use about one third of the energy of a typical new custom building. The connected lighting load for the building is a scant 0.77W/sq.ft.—over 46 percent less than ASHRAE/IESNA 90.1-2001 allowances. Continual energy monitoring has proven the benefit of the control strategy and actual daytime lighting energy use is over 40 percent less than the connected load, even in winter months.

Actual electricity usage by lighting was very close to expectations, consuming 40,000 kWh from March 2005 through February 2006. The building is one of only a handful of manufacturing facilities to earn a LEED Gold certification. NRG Systems also succeeded in its goal of using low cost energy and is supplying 72 percent of the energy used from renewable sources, which include solar, wind and wood pellets.

Company founders David and Jan Blittersdorf report having paid an 8.21 percent premium on the $7,833,000 project for the specialized design and construction, including a substantial investment in a photovoltaic system. According to David, "We have essentially prepaid our energy bill by relying on renewable energy and, a result, we won't have to worry about rising energy costs in the future." He estimates a $4-8 million saving over the 30-40 year projected life span of the building. Of the approximately $643,000 in premium building costs, $460,000 was for renewables and $183,000 ($3.93/sq. ft.) was the LEED cost premium. Designer Andy Shapiro concludes, "The cost premium for the daylighting design, coupled with carefully selected energy-efficient luminaires and controls, is rapidly repaid in miniscule utility bills."

Part IV

Controls for
Code Compliance

Chapter 9

Code-compliant Controls

Some designers consider them a necessary evil, others just another set of requirements to obtain a building permit. Energy codes are designed to set minimum standards for design and construction and can significantly reduce building system life-cycle costs.

U.S. energy codes address lighting prescriptively by setting lighting power density (LPD) limits on lighting for whole buildings. Setting LPD limits for whole buildings is important because energy-efficient lighting can be inefficient as a whole if installed in high densities in a building. Most codes now also mandate automatic lighting shutoff controls as well as controls for enclosed spaces.

This chapter provides general guidance to controls compliance in the most popular model commercial energy codes—the International Energy Conservation Code (IECC) and ASHRAE/IESNA Standard 90.1.

COMMERCIAL ENERGY CODES

Before 1992, states in the United States enacted energy codes on a voluntary basis, some developing their own codes while others adopted model codes. The Energy Policy Act of 1992 authorized DoE to establish a national energy standard for all states. Since 2004, Standard 90.1-1999, developed jointly between ASHRAE and IESNA, is the national energy standard. All states must have a commercial energy code at least as stringent as this.

However, the country is a patchwork of energy codes. Some states have adopted Standard 90.1-1999, while others have adopted the 2001 or 2004 versions (as of July 2007, a 2007 version is anticipated but not yet released). Many states have adopted the IECC. Often, these codes are adopted with amendments that individualize their codes. Other states, such as California, Florida, Oregon and Washington, have developed their own codes. And a few states have not complied with the DoE mandate. As of July 2007, the most common codes are IECC 2003, Standard 90.1-2004 and California's unique Title 24 code. Although there has been some convergence towards consistent requirements (IECC references Standard 90.1 as an alternative compliance standard), these differences can be confusing for design firms working in multiple jurisdictions.

Scope

ASHRAE 90.1 covers commercial buildings and residential buildings four stories and up above grade. This includes new buildings and their systems, new portions of buildings and their systems, and new systems and equipment in existing buildings. In an existing building, the lighting requirements of 90.1 go into effect if 50 percent or more of the light fixtures are replaced. The lighting controls requirements go into effect if the existing controls are replaced. If fewer than 50 percent of the fixtures are replaced, the code does not apply as long as the alterations do not increase the installed interior lighting power.

IECC covers commercial buildings or portions of commercial buildings undergoing new construction or additions, alterations, renovations or repairs.

COMMERCIAL CODES BY STATE

Table 9-1 provides a list of states and their commercial energy code requirements as of April 8, 2007. Confirm your state's latest status by visiting http://www.energycodes.gov/implement/state_codes/state_status_full.php and scrolling down to the commercial code section of the page.

Printed and bound by CPI Group (UK) Ltd, Croydon, CR0 4YY

23/10/2024

01777698-0004

Table 9-1. Status of commercial energy code adoption as of April 8, 2007. Source: DoE.

	Commercial Code	Enforcement Status	Approximate Stringency	Notes
Alaska	None	None Without Amendments	No Information	None statewide. All public facilities must be designed to comply with the thermal and lighting energy standards adopted by the Alaska Department of Transportation and Public Facilities under AS44.42.020(a)(14).
Alabama	None	Mandatory Without Amendments	As stringent as the ASHRAE 01	The Alabama Building Energy Conservation Code (ABECC) is a mandatory building code for state government buildings, administered by the Alabama Building Commission. The latest version of the Code (ABECC 2004), which is based on ASHRAE/IESNA 90.1 – 2001, was adopted in March 2005 and was implemented by the Alabama Building Commission in September 2005.
Arkansas	2003 IECC	Mandatory Without Amendments	As stringent as the 2003 IECC	ASHRAE/IESNA 90.1-2001, which is referenced by the 2003 IECC.
American Samoa	None	None Without Amendments	No Information	None.
Arizona	ASHRAE 99	Voluntary Without Amendments	As stringent as the ASHRAE 99	State-owned or -funded buildings, must comply with ASHRAE/IESNA 90.1-1999.
California	State Specific Code	Mandatory With Amendments	More stringent than the ASHRAE 04	State-developed code, Part 6 of Title 24, which meets or exceeds ASHRAE/IESNA 90.1-2004, is mandatory statewide as of Oct. 1, 2005.
Colorado	2003 IECC	Voluntary Without Amendments	As stringent as the 2003 IECC	Voluntary state provisions are based on 2003 IECC with reference to ASHRAE 90.1-2001
Connecticut	2003 IECC	Mandatory Without Amendments	As stringent as the 2003 IECC	2003 IECC with reference to ASHRAE 90.1-2001.
District of Columbia	2000 IECC	Mandatory Without Amendments	As stringent as the 2000 IECC	including reference to ASHRAE 90.1-1999
Delaware	ASHRAE 99	Mandatory Without Amendments	As stringent as the ASHRAE 99	ASHRAE 90.1-1999 provided that the respective county and municipality government shall exclude agricultural structures from the provisions.
Florida	State Specific Code	Mandatory With Amendments	More stringent than the ASHRAE 01	State-developed code, which meets or exceeds ASHRAE/IESNA 90.1-2001 is mandatory statewide.
Georgia	2000 IECC	Mandatory With Amendments	More stringent than the 2000 IECC	2000 IECC with Georgia State Amendments to include ASHRAE 90.1-2004 with Georgia Amendments became effective Jan. 1, 2006
Guam	ASHRAE 89	Mandatory Without Amendments	As stringent as the ASHRAE 89	ASHRAE/IESNA 90.1-1989.

Table 9-1 (*Continued*). Status of commercial energy code adoption as of April 8, 2007. Source: DoE.

Hawaii	None	Voluntary Without Amendments	No Information	Honolulu, Maui, and Kaui County require compliance with ASHRAE 90.1-1999. Hawaii County requires compliance with ASHRAE 90.1-1989.
Iowa	2006 IECC	Mandatory Without Amendments	As stringent as the 2006 IECC	2006 IECC with reference to ASHRAE 90.1-2004
Idaho	2003 IECC	Mandatory Without Amendments	As stringent as the 2003 IECC	2003 IECC
Illinois	2001 IECC	Mandatory Without Amendments	As stringent as the 2001 IECC	2000 IECC with the 01 Supplement
Indiana	State Specific Code	Mandatory With Amendments	stringent than the 90A90B	Indiana Energy Conservation Code (1992 Model Energy Code with Indiana amendments)
Kansas	2003 IECC	Mandatory Without Amendments	As stringent as the 2003 IECC	2003 IECC
Kentucky	2003 IECC	Mandatory With Amendments	As stringent as the 2003 IECC	
Louisiana	ASHRAE 01	Mandatory With Amendments	As stringent as the ASHRAE 01	No economizers are required.
Massachusetts	State Specific Code	Mandatory With Amendments	More stringent than the 2001 IECC	Elements from both the ASHRAE/IESNA 90.1-1999 and the International Energy Conservative Code (IECC), with state specific amendments.
Maryland	2003 IECC	Mandatory Without Amendments	As stringent as the 2003 IECC	
Maine	ASHRAE 01	Mandatory With Amendments	As stringent as the ASHRAE 01	ASHRAE/IESNA 90.1-2001
Michigan	ASHRAE 99	Mandatory With Amendments	As stringent as the ASHRAE 99	ASHRAE 90.1-1999 is the current standard. The new rules were effective March 13, 2003.
Minnesota	State Specific Code	Mandatory With Amendments	More stringent than the ASHRAE 89	Minnesota State Building Code, based on ASHRAE/IESNA 90.1-1989
Missouri	None	None Without Amendments	No Information	None, except state-owned buildings must comply with ASHRAE/IESNA 90.1-1989.
Commonwealth of the Northern Mariana Islands	State Specific Code	Mandatory Without Amendments	No Information	State-developed code, which adopts the 1991 Uniform Building Code is mandatory for all new and remodeled multi-family and commercial buildings.
Mississippi	None	None Without Amendments	No Information	90-1975 is mandatory for state-owned buildings, public buildings, and high-rise buildings only.
Montana	2003 IECC	Mandatory Without Amendments	As stringent as the 2003 IECC	2003 IECC with reference to ASHRAE 90.1-2001
North Carolina	State Specific Code	Mandatory With Amendments	More stringent than the 2000 IECC	State-developed code, modeled on the 2003 IECC with amendments including ASHRAE/IESNA 90.1-2004.
North Dakota	ASHRAE 89	Voluntary Without Amendments	As stringent as the ASHRAE 89	ASHRAE/IESNA 90.1-1989 is contingent on adoption by local jurisdiction

Table 9-1 (*Continued*). Status of commercial energy code adoption as of April 8, 2007. Source: DoE.

Nebraska	2003 IECC	Mandatory Without Amendments	As stringent as the 2003 IECC	2003 IECC with reference to ASHRAE 90.1-2001
New Hampshire	2000 IECC	Mandatory Without Amendments	As stringent as the 2000 IECC	2000 IECC with reference to ASHRAE 90.1-1999
New Jersey	ASHRAE 04	Mandatory With Amendments	As stringent as the ASHRAE 04	On 2/20/07 NJ adopted ASHRAE/IESNA 90.1-2004 with minor modifications. A six month interim period allows users to show compliance to NJ's previous code or current energy code. Previous code was based on ASHRAE 90.1-1999.
New Mexico	2003 IECC	Mandatory Without Amendments	As stringent as the 2003 IECC	July 1, 2004 IECC 2003 became effective.
Nevada	2003 IECC	Mandatory Without Amendments	As stringent as the 2003 IECC	
New York	2001 IECC	Mandatory With Amendments	As stringent as the 2001 IECC	2001 IECC w/amendments.
Ohio	ASHRAE 04	Mandatory Without Amendments	As stringent as the ASHRAE 04	ASHRAE 90.1-2004 became effective Sept. 6, 2005. Can show compliance to either 2003 IECC or 90.1-04.
Oklahoma	2003 IECC	Mandatory Without Amendments	As stringent as the 2003 IECC	2003 IECC is mandatory for jurisdictions without codes and for all state owned and leased facilities.
Oregon	State Specific Code	Mandatory With Amendments	More stringent than the ASHRAE 99	State-developed code that meets or exceeds ASHRAE/IESNA 90.1-1999 is mandatory statewide.
Pennsylvania	2006 IECC	Mandatory With Amendments	As stringent as the 2006 IECC	2006 IECC with reference to ASHRAE 90.1-2004
Puerto Rico	State Specific Code	Mandatory With Amendments	Less stringent than the ASHRAE 89	The Code for Energy Conservation in Puerto Rico, based on ASHRAE/IESNA 90.1-1989, is mandatory for the entire island of Puerto Rico.
Rhode Island	2003 IECC	Mandatory Without Amendments	As stringent as the 2003 IECC	With reference to ASHRAE 90.1-2001
South Carolina	2003 IECC	Mandatory Without Amendments	As stringent as the 2003 IECC	2003 IECC with reference to ASHRAE 90.1-2001
South Dakota	None	None Without Amendments	No Information	None.
Tennessee	90A90B	Voluntary Without Amendments	As stringent as the 90A90B	Local jurisdictions have the option of upgrading the energy efficiency code to 2000 IECC with 2001 amendments.
Texas	2001 IECC	Mandatory Without Amendments	As stringent as the 2001 IECC	2000 IECC with 2001 Supplement
Utah	2006 IECC	Mandatory Without Amendments	As stringent as the 2006 IECC	with reference to ASHRAE 90.1-2004
Virginia	2003 IECC	Mandatory With Amendments	As stringent as the 2003 IECC	2003 IECC with reference to ASHRAE 90.1-2004 effective November 2005
U.S. Virgin Islands	None	None Without Amendments	No Information	None.

Table 9-1 (*Continued*). Status of commercial energy code adoption as of April 8, 2007. Source: DoE.

Vermont	State Specific Code	Mandatory With Amendments	More stringent than the 2004 IECC	Based on 2004 IECC with amendments to include ASHRAE 90.1-2004
Washington	State Specific Code	Mandatory With Amendments	More stringent than the ASHRAE 01	State-developed code that meets or exceeds ASHRAE/IESNA 90.1-2001. Most recent updates effective July 1, 2007.
Wisconsin	State Specific Code	Mandatory With Amendments	As stringent as the 2000 IECC	2000 IECC w/amendments; can use COMcheck for building envelope, but not for HVAC or lighting. Set the code to be used with the "2000 IECC". Multi family buildings (3 stories or less, 3 dwellings or more) are considered commercial buildings in Wisconsin. REScheck may be used with these buildings if program is set for use with the "2000 IECC".
West Virginia	2003 IECC	Mandatory Without Amendments	As stringent as the 2003 IECC	
Wyoming	None	Voluntary Without Amendments	As stringent as the PRIOR 90A90B	The ICBO Uniform Building Code, which is based on the 1989 MEC, may be adopted and enforced by local jurisdictions.

ENERGY PROGRAMS

Two major programs use commercial energy codes as a standard to meet and exceed to earn incentives.

The Commercial Buildings Deduction's Interim Lighting Rule provides an accelerated tax deduction of $0.30-$0.60/sq.ft. for exceeding ASHRAE/IESNA Standard 90.1-2001 by 25-50 percent. For more information, visit www.lightingtaxdeduction.org.

As of July 2007, the LEED-NC v.2.2 green building rating system requires meeting ASHRAE/IESNA Standard 90.1-2004 as a minimum prerequisite, and awards points based on exceeding the Standard (as of July 2007, at least 2 Energy and Atmosphere points are required for most LEED projects). For more information, visit www.usgbc.org.

CALIFORNIA'S TITLE 24

California has what many consider the strictest energy code in the country—*The Energy Efficiency Standards for Residential and*

Nonresidential Buildings, Title 24, Part 6, of the California Code of Regulations. The most recent version, which took effect October 1, 2005, included the following changes.

- Deleted unconditioned space exemption
- Added control certifications
- New skylight and daylighting requirements
- Further reductions in lighting power
- Revised tailored approach for retail spaces
- Added outdoor lighting zones, energy calculations and controls
- Added sign control
- Added residential mandates

Title 24 is not specifically covered further in this chapter. For more information, visit www.energy.ca.gov/title24.

CODE SNAPSHOT

With various rules and exceptions, IECC and 90.1 present the lighting requirements shown in Table 9-2. A basic compliance map is shown in Figure 9-1.

CODE COMPLIANCE

How often is the code actually complied with? How is the code enforced and who enforces it? Who on the design team carries the most responsibility and decision-making authority related to compliance? What are the most significant barriers to compliance?

A 2007 study conducted by ZING Communications, Inc., sponsored by Architectural Products Magazine and the Lighting Controls Association, attempted to answer these questions. The study was based on a survey distributed to 11,000 commercial architects, electrical engineers, lighting designers and building contractors. The results suggest:

Table 9-2. Snapshot of popular model commercial energy codes.

	IECC		ASHRAE/IESNA 90.1		
	2003	2006	1999	2001	2004
Automatic Lighting Shutoff					
Scheduling OR	x	x	x	x	x
Occupancy sensors OR	x	x	x	x	x
Another automatic method	x	x	x	x	x
Space Controls					
Manual OR	x	x	x	x	x
Automatic (including an occupancy sensor) AND			x	x	x
Employee lunch/break room, higher education classroom and conference/meeting room lighting must be activated by either multi-scene control or occupancy sensor					x
Guest Rooms/Sleeping Units (Hotels, Motels, etc.)					
Must have control at room entrances to turn off permanently installed and switched receptacle lighting	x	x	x	x	x
Display/Accent, Non-Visual (Such as for Plant Growth), Demonstration Lighting					
Separate, independent control required			x	x	x
Task Lighting					
Integral control device or readily accessible wall-mounted device			x	x	x
Light Level Reduction					
Manual switch enabling bi-level switching OR	x	x			
Dimming OR	x	x			
Occupancy sensors OR	x	x			
Daylight harvesting control	x	x			
Exterior Lighting Control					
Photosensor OR	x	x	x	x	x
Astronomical time switch	x	x	x	x	x
Tandem Wiring					
Tandem wiring requirements	x	x	x	x	x
Exit Signs					
· Minimum source efficacy of 25 LPW			x	x	
No more than 5W per face	x	x			x
Exterior Building Grounds Lighting (other than low-voltage landscape fixtures)					
Fixture >100W must have source efficacy of 45 LPW -	x				
Fixture >100W must have source efficacy of 60 LPW OR		x	x	x	x
Fixture >100W must be controlled by motion sensor		x	x	x	x
Interior Lighting Power					
Building method OR	x	x	x	x	x
Space by space method OR	x	x	x	x	x
Performance method	x	x	x	x	x
Exterior Lighting Power: Prescriptive LPD allowances					
Exterior lighting LPD allowances		x	x	x	x
Electrical Energy Consumption					
If building contains individual dwelling units, each unit must be separately metered	x	x			

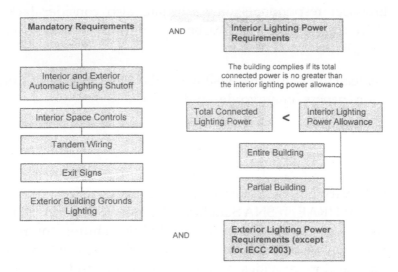

Figure 9-1. Basic compliance map for IECC and Standard 90.1.

- An estimated 80 percent compliance rate as a weighted average across all respondents.

- Jurisdictions are more likely than not to require documentation or intent to comply with the applicable commercial energy code as a prerequisite for obtaining a commercial building permit.

- It is more common that the organization with the authority to interpret the code, approve its application, and then inspect the project to verify compliance, is the local building department—specifically, an individual who also handles structural, plumbing, etc.

- About one in 10 respondents reports that compliance inspections do not occur in their jurisdictions.

- Value engineering, or a focus on initial cost that can result in the removal of critical lighting choices, is the most significant barrier to code compliance.

- Engineer respondents, in particular, also consider lack of strict code enforcement to be a significant barrier to code compliance.

To download this study free, visit www.aboutlightingcontrols.org/education/papers/energycode_study.html.

CONTROLS FOR COMPLIANCE

With varying exceptions and requirements, IECC 2003 and 2006 and ASHRAE/IESNA Standard 90.1-1999, -2001 and -2004 all require interior and exterior automatic lighting shutoff controls as well as manual or automatic controls in interior enclosed spaces.

This section provides a code comparison and a guide to compliant controls options for the following controls provisions:

- Interior automatic lighting shutoff (IECC, 90.1)
- Interior space controls (IECC, 90.1)
- Interior light level reduction control (IECC)
- Display/Accent lighting control (90.1)
- Exterior lighting control (IECC, 90.1)

Several other controls provisions, including independent controls for non-visual lighting (such as for plant growth), demonstration lighting (lighting for sale or part of lighting education), and task lighting, are not covered further in this section.

By the end of this section, you will have learned major controls provisions in IECC 2003/2006 and Standard 90.1-1999/2001/2004 and what control strategies may be used to comply with code requirements.

Automatic Lighting Shutoff

With varying exceptions and rules, Standard 90.1 and IECC require automatic lighting shutoff to ensure that general lighting in a building is turned off when it is no longer needed. Three methods of compliance of interest include:

- Scheduled shutoff
- Occupancy-based shutoff
- Timed shutoff

Table 9-3. Summary comparison of automatic lighting shutoff requirements.

	IECC 2003	IECC 2006	90.1-1999	90.1-2001	90.1-2004
Requirement	Mandatory				
Application	Interior lighting in buildings >5,000sq.ft.				
Exceptions	- Areas with only 1 light fixture - Corridors - Storerooms - Restrooms - Public lobbies - Dwelling units	- Sleeping units - Areas directly involving patient care - Areas where shutoff would endanger safety or security	- Lighting intended for 24-hour operation - Emergency lighting automatically off during normal building operation - Lighting in living units - Lighting specifically required by life safety law or regulation - Decorative gas lighting	- Lighting intended for 24-hour operation - Emergency lighting automatically off during normal building operation - Lighting in living units - Lighting specifically required by life safety law or regulation - Decorative gas lighting	- Lighting intended for 24-hour operation - Emergency lighting automatically off during normal building operation - Lighting in living units - Lighting specifically required by life safety law or regulation - Decorative gas lighting - Used for patient care - Where shutoff would endanger safety or security
Acceptable Methods	- Occupancy sensors - Time-of-day scheduling device - Occupant intervention on unscheduled basis (intended for limited applications)	- Occupancy sensor - Time-of-day scheduling device - Signal from another control or alarm system such as BAS	- Occupancy sensor - Time-of-day scheduling device - Occupant intervention on unscheduled basis (intended for limited applications)	- Occupancy sensor - Time-of-day scheduling device - Signal from another control or alarm system such as BAS	- Occupancy sensor - Time-of-day scheduling device - Signal from another control or alarm system such as BAS
Occupancy Sensors	No special requirements	Must turn lights off within 30 minutes of occupant leaving space	Must turn lights off within 30 minutes of occupant leaving space	Must turn lights off within 30 minutes of occupant leaving space	Must turn lights off within 30 minutes of occupant leaving space
Scheduling	Must have independent program that controls lights in areas of 25,000 sq.ft. maximum size and are not more than 1 floor; some building types must incorporate automatic holiday schedule	Must have independent program that controls lights in areas of 25,000 sq.ft. maximum size and are not more than 1 floor	Must have independent program that controls lights in areas of 25,000 sq.ft. maximum size and are not more than 1 floor	Must have independent program that controls lights in areas of 25,000 sq.ft. maximum size and are not more than 1 floor	Must have independent program that controls lights in areas of 25,000 sq.ft. maximum size and are not more than 1 floor

Scheduled Shutoff

Scheduling entails activating and deactivating controlled lighting loads automatically based on a schedule, usually normal business operating hours. It is a proven energy management strategy for which savings of 5-15 percent have been demonstrated. Scheduling can be achieved by using a lighting control panel that features a time-clock.

Scheduling is a time-based function and as a consequence it is most suited for facilities or spaces where the occupancy pattern is predictable and certain things happen at certain times. Typical applications include open offices, retail sales floors, hallways and common areas.

If the building needs more than one lighting control panel, one approach is to centralize the time clock in a single panel (master) and then data-network the panel to other panels in the system (slaves) to communicate scheduled events. Another is to decentralize the scheduling function in a series of panels connected to a common bus, with the benefit being a failure in a single point in the system will not affect the rest of the system. For centralized programming and monitoring, consider a system that operates on a PC with lighting control software.

Because "off-normal" conditions inevitably arise, codes require capability for occupants to override the shutoff and turn on local lighting for use, covered later in this chapter.

Occupancy-based Shutoff

Occupancy-based automatic switching entails using occupancy sensors to shut off controlled lighting loads automatically when these devices detect the absence of people. It is a proven energy management strategy for which savings of 10-50+ percent have been demonstrated. Typical applications include private offices, restrooms, classrooms, conference rooms, break rooms, etc. Occupancy sensors are ideally suited for spaces:

*   In which the lighting is not required to be operating all day for safety or security reasons. For example, occupancy sensors

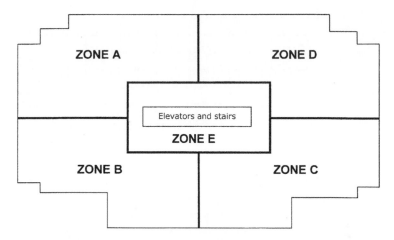

ZONE A

ZONE D

Elevators and stairs

ZONE E

ZONE B

ZONE C

Figure 9-2. The general lighting in this 10,000-sq.ft. floor of an office building can be controlled by an intelligent control panel as a single zone with its own programmed schedule per IECC and Standard 90.1. However, code override provisions require the ability for users to override the shutoff and turn on the lights in a maximum area—in this case, 2,500 sq.ft. per Standard 90.1—in turn requiring more granular zoning.

are not recommended for public spaces such as hallways and lobbies, where the lights must remain on even when the space is unoccupied

- That are intermittently occupied throughout the day or are otherwise left vacant for significant portions of the day, and where occupancy is less predictable

- Smaller projects

- Projects requiring more granular control

Timed Shutoff

Timer switches turn off the lights in a single load switch leg after a preset period of time once the lights have been switched on. These switches may be programmable electronic switches or spring-loaded, mechanical, twist-timer switches. The shut-off

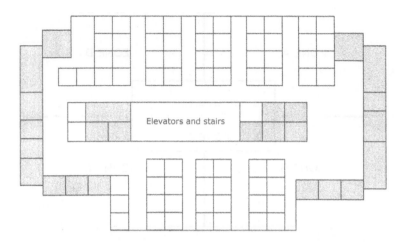

Figure 9-3. This 10,000-sq.ft. office building footprint delineates conference rooms, private offices, a kitchen and a copy room (gray) from open office cubicle and other spaces (white). Occupancy in these spaces is intermittent and unpredictable, making them ideal opportunities for occupancy-based shutoff throughout the day.

setting is determined by the user or the contractor, depending on the technology.

For example, a user enters a room, activates the lights and the timer, and when the timer expires, the lights shut off. When the lights are about to shut off, a warning signal may be emitted. This makes the timer switch both an occupancy- and time-based strategy to save energy that can be available for less than one-third the cost of an occupancy sensor.

However, timer switches typically save less energy than occupancy sensors, and may experience nuisance switching, as the lights will shut off at the end of the period unless the user restarts the timer. For this reason, timer switches are typically used in storage rooms, mechanical and electrical rooms, supply closets and janitorial spaces.

Lighting Controls in Enclosed Spaces
With varying exceptions and rules, codes require additional controls in spaces enclosed by ceiling-height partitions. This enables

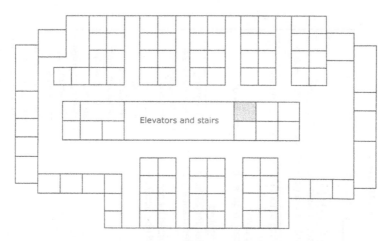

Figure 9-4. Small spaces with limited use, such as the utility room shaded gray in the above office building footprint, are ideal applications for timed shutoff.

users to override automatic shutoff events for a period of time while activating only localized lighting. Depending on the space and code in effect, this may be accomplished using manual switches, occupancy sensors or multi-scene controls such as dimmers. Two methods of compliance of interest include:

- Lighting control panels with timed override switches
- Occupancy sensors

Control Panel with Timed Override Switches
 As stated above, lighting control panels are a highly suitable automatic lighting shutoff strategy for spaces with high usage and regular schedules.
 Panels can be networked with automatic control switches to provide local override capability to users. These switches are essentially manual ON/OFF switches. However, because the shutoff override is limited to 2-4 hours or longer depending on the code in effect, these switches can also receive signals from the control panel to turn controlled lighting loads on or off—enabling a timed override.

Table 9-4. Summary comparison of space control code requirements.

	IECC 2003	IECC 2006	90.1-1999	90.1-2001	90.1-2004
Requirement			Mandatory		
Application		Building interior spaces enclosed by ceiling-height partitions			
Requirement	The space will have at least one manual control controlling the lights in the space.		The space will have at least one control device that independently controls the general lighting in the space.		
Exceptions	- Spaces that must be continuously lighted for safety or security - Stairways and corridors that are part of the means of egress - Lighting within dwelling units		- Emergency lighting automatically off during normal building operation - Lighting in living units - Lighting specifically required by life safety law or regulation - Decorative gas lighting		
Acceptable Methods	- Manual control located so that user can see lights being controlled - Remote manual control with controlled lights annunciated - If the space is a "guestroom" (2003) or "sleeping unit" (2006) in a hotel, motel, boarding house or similar building, master switch must be provided at entry door that controls all permanently installed fixtures and switched receptacles except bathroom(s); if the space is a suite, such a control will be provided at entry to each room or at suite's primary entry		- If lighting intended for 24-hour operation, activated manually by occupant - Otherwise, activated by occupancy sensor or manually by occupant		- If lighting intended for 24-hour operation, activated manually by occupant - If space is classroom (not including shop, lab and preschool-grade 12 classroom), conference/ meeting room or employee lunch/ break room, must be activated by occupancy

	sensor (with max. 30-min. time delay) or multi-scene control (such as dimmer) activated by occupant - Otherwise, activated by occupancy sensor or manually by occupant	- If maximum area of enclosed space is ≤10,000 sq.ft., then the control's maximum coverage area is 2,500 sq.ft. - If maximum area of enclosed space is >10,000 sq.ft., then the control's maximum coverage area is 10,000 sq.ft. - Control must be readily accessible and located so user can see controlled lights - Control can be remote for safety or security reasons, but controlled lighting must be annunciated - Control cannot override time-scheduled automatic shutoff for more than 4 hours
Special requirements	- Control cannot override time-scheduled automatic shutoff for more than 2 hours - Area controlled may not exceed 5,000 sq.ft. - If building/space a mall, arcade, auditorium, single-tenant retail space, industrial space or arena where captive-key override is utilized, then override can exceed 2 hours and controlled area may not exceed 20,000 sq.ft.	

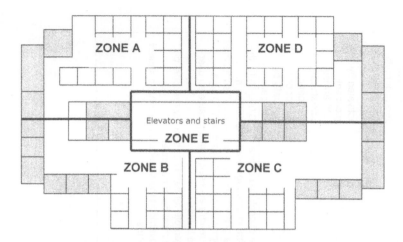

Figure 9-5. The above 10,000-sq.ft. floor of an office building utilizes a combination of scheduling and timed override switches for control of open office areas and public spaces. After a maximum period of time allowed by code, the switches receive a signal from the control panel to turn the lights off. Per Standard 90.1, the maximum space control zone for a space this size is 2,500 sq.ft., leading to its division into four primary zones and a fifth zone centered on the elevator lobby and area around the stairs because this space has a different use. (IECC recognizes maximum coverage for each control as 5,000 sq.ft. for this building/space type, meaning there could be three zones in this floor.) Meanwhile, the gray-shaded areas represent enclosed spaces in which other controls choices can be adopted.

Occupancy Sensors

As stated above, occupancy sensors are a highly suitable automatic lighting shutoff strategy for spaces that are intermittently occupied and left unoccupied for periods of time throughout the day.

One advantage of occupancy sensors is that they contain their own override (they turn off the lights when the space is unoccupied for a designated period of time). They are particularly suitable for smaller enclosed spaces, which require a high degree of control granularity or resolution.

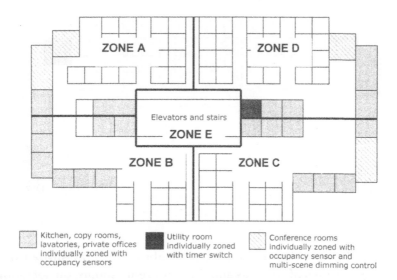

Kitchen, copy rooms, lavatories, private offices individually zoned with occupancy sensors

Utility room individually zoned with timer switch

Conference rooms individually zoned with occupancy sensor and multi-scene dimming control

Figure 9-6. The kitchen, copy room and private offices are intermittently occupied, left vacant for significant portions of the day, and represent a highly granular group of control zones, making them highly suited for occupancy sensors both for shutoff and override. The conference rooms in this building utilize multi-scene dimming controls in addition to occupancy sensors.

Light Level Reduction Control

With several exceptions, IECC 2003 and 2006 require that occupants be able to reduce lighting load in the space in a reasonably uniform pattern by at least 50 percent (see Table 9-5). There is a notable exception for spaces controlled by occupancy sensors.

IECC recognizes four methods of light level reduction control:

- Controlling all lamps or fixtures (e.g., dimming)
- Dual switching alternate rows, fixtures or lamps
- Switching middle lamp independent of outer lamps (3-lamp fixtures)
- Switching each fixture or each lamp

Suitable solutions include dimming controls, manual switches and daylighting controls.

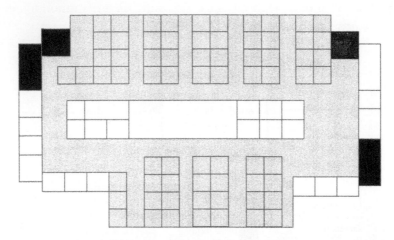

Figure 9-7. In our office building example above, the gray-shaded areas, including the open office areas and the conference rooms, do not qualify for exceptions and therefore must adopt some form of light level reduction control (in this case, an additional override switch that controls alternate lamps in the ceiling lighting system, matching the control panel and override zones). The conference rooms, shown as black-shaded areas, are covered not only because occupancy sensors qualify these space as exceptions, but because multi-scene dimming controls are used, which qualify as a suitable light level reduction control.

Display/Accent Lighting Control

Standard 90.1-1999, -2001 and -2004 require that display/accent lighting be controlled independently of the general lighting. This can be accomplished, for example, with lighting control panels and programmable switches.

By separately controlling display/accent lighting from general lighting in a store, for example, display lighting can be scheduled to operate only during shopping hours, while the general lighting can operate during the store's full operating hours.

Exterior Lighting Control

With varying exceptions and rules, codes require that exterior lighting be automated via an astronomical time switch, photosensor, or both.

Table 9-5. Summary comparison of light level reduction code requirements.

	IECC 2003	IECC 2006	90.1-1999	90.1-2001	90.1-2004
Requirement	Mandatory				
Application	Each interior space enclosed by ceiling-height partitions				
Requirement	Occupants must be able to reduce lighting load in reasonable uniform pattern by at least 50%				
Exemptions	Spaces with occupancy sensors; stairways and corridors that are part of means of egress; security or emergency area that must be continuously lighted; spaces with only one light fixture; corridors, restrooms, storerooms and public lobbies; guestrooms (2003)/sleeping units (2006), spaces with a lighting power density <0.6W/sq.ft.		No requirements		
Acceptable Methods	Controlling all lamps or fixtures (e.g., dimming); dual switching of alternate rows, fixtures or lamps; switching middle lamp independent of outer lamps; switching each fixture or lamp				

Astronomical Time Control

Astronomical time control can be gained using a lighting control panel with an astronomical time clock to implement an ON/OFF schedule. The astronomical time clock uses location and time data to ensure that the lighting can be switched in sync with the changing seasons.

While it can be used for dusk-to-dawn lighting depending on the code in effect, it is particularly effective for turning off non-security lighting earlier in the night. Suitable applications include parking lots, walkways and building facades.

Photosensor Control

Photosensors or photocells are devices that measure available daylight and then switch connected lighting loads on when there is sufficient daylight at dawn, and off when daylight declines sufficiently at dusk. As a result, photo sensors are effective for dusk to dawn fixtures, such as security lighting, but not lighting that is turned off during the night because it is no longer needed. Often, the photosensor is mounted on the roof of the building.

Suitable applications include parking lots, walkways, building facades and security lighting.

Table 9-6. Summary comparison of exterior lighting control code requirements.

	IECC 2003	IECC 2006	90.1-1999	90.1-2001	90.1-2004
Requirement	Mandatory				
Application	All exterior lighting				
	Exterior lighting must be automatically shut off when either 1) sufficient daylight is available or 2) when not required during nighttime hours				
Exceptions	Lighting intended for 24-hour operation	Exterior lighting for covered vehicle entrances or exits from buildings or parking structures where required for safety, security or eye adaptation	- Exterior lighting for covered vehicle entrances or exits from buildings or parking structures where required for safety, security or eye adaptation - Emergency lighting automatically off during normal building operation - Lighting specifically required by life safety law or regulation - Decorative gas lighting - Specialized signal, directional or marker lighting equipped with independent control device - Lighting with independent control device and used to highlight features of public monuments or registered historical landmark structures or buildings - Lighting with independent control device and integral to advertising signage		- Exterior lighting for covered vehicle entrances or exits from buildings or parking structures where required for safety, security or eye adaptation - Emergency lighting automatically off during normal building operation - Lighting specifically required by life safety law or regulation - Decorative gas lighting
Acceptable Methods	Astronomical time switch OR photosensor (dusk to dawn lighting operation)				
Special Requirements	If using automatic switching control, must have combination 7-day and seasonal daylight program schedule adjustment, and minimum 4-hour power backup	If using astronomical time switch, must be capable of retaining programming and time setting during power outage for at least 10 hours	None		If lighting is not for dusk to dawn operation, then MUST use an astronomical time switch using astronomical time switch capable of retaining programming and time setting during power outage for at least 10 hours
Additional Mandatory Requirement for Building Grounds Lighting	All exterior building grounds light fixtures operating >100W must have a minimum efficacy of 45 lumens/W—IF these fixtures receive power via building's energy	All exterior building grounds light fixtures operating >100W must have a minimum efficacy of 60 lumens/W OR be either controlled by a motion sensor— IF these fixtures	All exterior building grounds light fixtures operating >100W must be either controlled by a motion sensor OR have a minimum efficacy of 60 lumens/W		

(Continued)

Table 9-6 (*Continued*). **Summary comparison of exterior lighting control code requirements.**

	IECC 2003	IECC 2006	90.1-1999	90.1-2001	90.1-2004
Requirement	Mandatory				
Application	All exterior lighting				
	Exterior lighting must be automatically shut off when either 1) sufficient daylight is available or 2) when not required during nighttime hours				
	service	receive power via building's energy service			
Exceptions to Additional Mandatory Requirement for Building Grounds Lighting	- Low-voltage landscape lighting - Where approved because of historical, safety, signage or emergency considerations		- Emergency lighting automatically off during normal building operation - Lighting specifically required by life safety law or regulation - Decorative gas lighting - Specialized signal, directional or marker lighting associated with transportation - Lighting that is integral to advertising signage - Lighting used to highlight features of public monuments and registered historic landmark structures or buildings		- Emergency lighting automatically off during normal building operation - Lighting specifically required by life safety law or regulation - Decorative gas lighting - Specialized signal, directional or marker lighting associated with transportation - Advertising or directional signage - Lighting integral to equipment or instrumentation and installed by its manufacturer - Lighting for theatrical purposes (e.g., performance, stage, film/TV production) - Temporary lighting. - Lighting for athletic playing areas - Lighting for industrial production, material handling, transportation sites and associated storage areas - Theme elements in a theme/amusement park - Lighting used to highlight features of public monuments and registered historic landmark structures or buildings

Part V

Commissioning

Chapter 10

Commissioning

Commissioning entails systematically testing all controls in the building to ensure that they provide specified performance and interact properly as a system, so as to satisfy the design intent and owner's needs. It is vital to ensure proper operation, user acceptance, and energy savings potential in new construction as well as renovation and retrofit projects. Lack of commissioning is a leading cause of failure in lighting control projects.

Commissioning is now required in California construction projects with the implementation of the 2005 Title 24 energy code. The code requires that all controls become certified to meet minimum performance requirements. Further, the code requires formal acceptance and certification of projects to ensure controls are operating as required. Commissioning can also both required and can contribute to receiving credit in the LEED rating system.

Commissioning may involve all members of the team and is typically led by the commissioning agent, who may be a contractor, commissioning specialist, a manufacturer or other professional. Some projects may require that an independent third party such as an architectural or engineering firm perform commissioning.

Be sure to plan and budget for commissioning as part of the design and construction process. To encourage it, engineers should write detailed commissioning requirements in their project specifications. For example, specifying that the manufacturer will be perform a site startup and user training can be an effective means of ensuring that its products are installed and operate as intended. Be sure not to delete commissioning as part of any project cost-saving measure.

Commissioning is differentiated from factory start-up, which occurs prior to commissioning and entails the manufacturer or its representative ensuring that its products perform as intended within the designed system.

Calibration is a standard tool used in commissioning. Calibration means adjusting sensors to gain desired performance in the actual conditions of the application. Since sensors are used to provide information to the control system, which responds to that information, proper calibration of sensors is essential to get desired performance from the entire control system. In older systems, sensors are calibrated physically. In sophisticated systems, sensors may be calibrated using software. Calibration is essential for occupancy sensors to adjust factory default settings to application need. It is also important for photosensors. For example, a photosensor in a room with light-painted walls will respond differently than a photosensor in a room right next door with dark-painted walls.

After commissioning is completed, tell the users about the intent and functionality of the controls, especially the overrides. This is critical because if users do not understand the controls, they may bypass them. Give all documentation and instructions to the owner's maintenance personnel so that they can maintain and re-tune the system as needed.

Table 10-1 and Table 10-2 provide basic guidance on commissioning fluorescent switching and dimming systems. For more information about commissioning occupancy sensors, see Chapter 1.

Table 10-1. Commissioning and calibration activities for switching systems.

Control Type	Commissioning and Calibration
Occupancy sensors and	Ensure that the sensor is correctly placed and oriented per the photosensors specifications and/or construction drawings. If unanticipated obstructions are present, it may be necessary to adjust the sensor location and orientation.
Occupancy sensors	Adjust the sensitivity and time delay of the occupancy sensor, and test to ensure it provides appropriate response. For optimal energy savings and lamp life, set the time delay for a minimum of 15 minutes.
Daylight harvesting	All furnishings and interior finishes and materials should be installed before calibrating the sensors. Adjust the photosensor to determine the threshold for switching based on detected light level. It may be helpful to calibrate under normal daylight conditions and dusk conditions (it may be possible to close window blinds to approximate dusk). Record the calibration adjustments if possible and replicate in similar spaces.
Automatic shut-off ("sweep off")	Input the schedule into the programmable scheduling controls, incorporating weekday, weekend and holiday operating times. Ensure that overrides work and that they are located conveniently for users.

Table 10-2. Commissioning and calibration activities for fluorescent dimming systems.

Control Type	Calibration and Commissioning Activities
All	Ensure that sensors are properly located and oriented per specifications and/or construction drawings.
Dimming systems	Season new lamps by operating them continuously for a recommended period of time (see next page).
Daylight harvesting	All furnishings and interior finishes and materials should be installed before calibrating the sensors. Adjust the photosensor to determine the threshold for switching based on detected light level. It may be helpful to calibrate under normal daylight conditions and dusk conditions (it may be possible to close window blinds to approximate dusk). Record the calibration adjustments if possible and replicate in similar spaces.
Manual dimming	Ensure correct placement of the dimmer adjacent to the wall switch per the construction drawings. Adjust the upper limit of the dimming range according to the task being performed, and set the lower limit of the range so that the minimum level meets the use/application of the space.

Index